To my father, Michael Harris,

1922–2014

RIGOR MORTIS

RIGOR MORTIS

How Sloppy Science Creates Worthless Cures, Crushes Hope, and Wastes Billions

Richard Harris

BASIC BOOKS
New York

Published by Basic Books, an imprint of Perseus Books, LLC, a subsidiary of Hachette Book Group, Inc.

Books published by Basic Books are available at special discounts for bulk purchases in the United States by corporations, institutions, and other organizations. For more information, please contact the Special Markets Department at Perseus Books, 2300 Chestnut Street, Suite 200, Philadelphia, PA 19103, or call (800) 810-4145, ext. 5000, or e-mail special.markets@perseusbooks.com.

Designed by Jeff Williams

A catalog record is available from the Library of Congress.
ISBN: 978-0-465-09790-6 (hardcover)
ISBN: 978-0-465-09791-3 (e-book)

10 9 8 7 6 5 4 3 2 1

CONTENTS

PREFACE

WHEN YOU READ about advances in medicine, it often seems like long-awaited breakthroughs are just around the corner for cancer, Alzheimer's, stroke, osteoarthritis, and countless less common diseases. But it turns out we live in a world with an awful lot of corners. Most of the time we round one only to discover another corner rather than a destination. I've reported countless medical stories since I arrived at National Public Radio in 1986, in retrospect with more hopefulness than they often deserved. And lately I've come to realize that the reason medical research is so painfully slow isn't simply because it's hard—which indeed it is. It also turns out that scientists have been taking shortcuts around the methods they are supposed to use to avoid fooling themselves. The consequences are now haunting biomedical research. Simply too much of what's published is wrong. It doesn't have to be that way.

These should be halcyon days for medical science. The human genome, our genetic blueprint, was deciphered and laid out for all to see in 2003. Technology for research labs has progressed at an astonishing pace. What used to take years of toil by a dedicated team can now be accomplished in an afternoon by a technician with the right instruments. Scientists can custom-design mice and engineer them to stand in for humans in laboratory experiments. Researchers can sift through terabytes of data to find clues for new diagnostic tests, treatments, and cures. To be sure, there have been great strides in medicine when viewed over the long haul—antibiotics, vaccines, and heart surgery, along with potent public health advice (especially, "Don't smoke!"). Life expectancy in the United States continues to creep upward, by and large. Underlying this ongoing effort is a generous pool of money. American taxpayers contribute more than $30 billion a year to fund the National Institutes of Health. Add in other sources, like the price of research that's baked into the pills we swallow and the medical treatments we receive, and the average American household spends $900 a year to support biomedical studies.

Yet metastatic cancer is nearly as unstoppable now as it was decades ago (with only a few exceptions). Alzheimer's disease remains untreatable, even as an avalanche of baby boomers ages and becomes more vulnerable to that grim and costly condition. Lou Gehrig's disease (amyotrophic lateral sclerosis, or ALS) is one of many devastating neurological conditions for which there is no effective remedy.

In fact, of the 7,000 known diseases, only about 500 have treatments, many offering just marginal benefits. As Malcolm Macleod at the University of Edinburgh put it, medical science is in the doldrums.

Biomedical science hasn't ground to a halt. Far from it. But this wasted effort is slowing progress—and at a time we can least afford it. After long periods of growth, federal support for biomedical research is now shrinking, given the growing cost of doing science. So it's never been more important to make the most of these precious resources.

Despite the technology, the effort, the money, and, yes, even the passion on the part of many scientists who are determined to make a difference, medical research is plagued with unforced and unnecessary errors. Scientists often face a stark choice: they can do what's best for medical advancement by adhering to the rigorous standards of science, or they can do what they perceive is necessary to maintain a career in the hypercompetitive environment of academic research. It's a choice nobody should have to make.

In the following chapters, you will read about the many ways that research has gone astray, as perverse incentives discourage scientists from following the rigorous path of top-quality science. My use of the term "rigor mortis," or the stiffness that comes after death, is of course a bit of hyperbole in the service of wordplay. Rigor in biomedical science certainly isn't dead, but it does need a major jolt of energy.

The good news is that these problems are now being recognized. Some can be fixed without a lot of technical difficulty. For example, any scientist can send a sample of cells off to a lab that can authenticate its identity. Researchers working at the lab bench can make small adjustments to their experiments to reduce the risk that wishful thinking is tainting their results. And biostatisticians can help make sure an experiment is designed and analyzed properly. The challenge now isn't identifying the technical fixes. The much harder challenge is changing the culture and the structure of biomedicine so that scientists don't have to choose between doing it right and keeping their labs and careers afloat.

In researching this book, I expected scientists would be reluctant to talk about the troubles facing their enterprise. That was surprisingly not the case. In fact, most scientists I called or visited were eager to tell their stories and to share their suggestions for how to put things right. Leaders at the National Institutes of Health and elsewhere have also stepped up to acknowledge these problems and seek solutions. (I've noticed that people are generally more willing to admit a problem exists when there are some concrete solutions at hand.) Patient-advocacy groups are increasingly addressing these problems. And a few pioneers have taken on these issues as their personal crusade. They know that medical science has already done much to serve humanity, and it has the potential to do so much more.

Let me end this introduction with an important philosophical point, relevant both to science and to this book. Most of science is built on inference rather than direct observation. We can't see the atoms or molecules inside our bodies, and we can't truly explain the root cause of disease. Science progresses by testing ideas indirectly, throwing out the ones that seem wrong, and building on those best supported by the facts at hand. Gradually, scientists build stories that do a better job of approximating the truth. But at any given moment, there are parallel narratives, sometimes sharply at odds with one another. Scientists rely on their own individual judgments to decide which stories come closer to the truth (absolute Truth is forever out of reach). Some stories that seem on the fringe today will become the accepted narrative some years from now. Indeed, it's the unexpected ideas that often propel science forward. Writers often don't say so clearly, but we too are in the business of weighing evidence and making judgment calls, assembling observations that bring us closer to the truth as we perceive it. It's a necessary element of storytelling. No doubt there will be those who see the world differently, who weigh a somewhat different set of facts and come to different conclusions. Since I explore the contingent nature of science in this book, it seems only fair to acknowledge that I'm making judgments as well, not revealing the objective Truth.

Chapter One

BEGLEY'S BOMBSHELL

IT WAS ONE of those things that everybody knew but was too polite to say. Each year about a million biomedical studies are published in the scientific literature. And many of them are simply wrong. Set aside the voice-of-God prose, the fancy statistics, and the peer review process, which is supposed to weed out the weak and errant. Lots of this stuff just doesn't stand up to scrutiny. Sometimes it's because a scientist is exploring the precarious edge of knowledge. Sometimes the scientist has unconsciously willed the data to tell a story that's not in fact true. Occasionally there is outright fraud. But a large share of what gets published is wrong.

C. Glenn Begley decided to say what most other people dared not speak. An Australian-born scientist, he had left academia after twenty-five years in the lab to head up cancer research at the pioneering biotech company Amgen in Southern California. While working in academia, Begley had codiscovered a protein called human G-CSF, which is now used in cancer treatments to reconstitute a person's immune system after a potentially lethal dose of chemotherapy. G-CSF ultimately proved to be Amgen's first blockbuster drug, so it's not surprising that years later, when the company wanted to create an entire cancer research program, it hired Begley for the job.

Pharmaceutical companies rely heavily on published research from academic labs, which are largely funded by taxpayers, to get ideas for new drugs. Companies can then seize upon those ideas, develop them, and make them available as new treatments. Begley's staff scoured the biomedical literature for hot leads for potential new drugs. Every time something looked promising, they'd start a dossier on the project. Begley insisted that the first step of any research project would be to have company scientists repeat the experiment to see if they could come up with the same results. Most of the time, Amgen labs couldn't. That was duly noted on the dossier, the case was closed, and the scientists moved on to the next exciting idea reported in the scientific literature.

After ten years at Amgen, Begley was ready to move on. But before he went, he wanted to take stock of the studies that his team had filed away as not reproducible—focusing in particular on the ones that could have led to important drugs, if they had panned out. He chose fifty-three papers he considered potentially groundbreaking. And for this review, the company didn't simply try to repeat the experiments—Begley asked the scientists who originally published these exciting results to help.

"The vast majority of the time the scientists were willing to work with us. There were only a couple of occasions where truly the scientists hung up on us and refused to continue the conversation," he said. First, Begley asked the scientists to provide the exact materials that they had used in the original experiment. If Amgen again couldn't reproduce the result with this material, they kept trying. "On about twenty occasions we actually sent [company] scientists to the host laboratory and watched them perform experiments themselves," Begley told me. This time, however, the original researchers were kept in the dark about which part of the experiment was supposed to produce positive results and which would serve as a comparison group (the control). Most of the time, the experiments failed under these blinded conditions. "So it wasn't just that Amgen was unable to reproduce them," Begley said. "What was more shocking to me was the original investigators themselves were not able to." Of the fifty-three original exciting studies that Begley put to the test, he could reproduce just six. Six. That's barely one out of ten.

Begley went to the Amgen board of directors and asked what he should do with this information. They told him to publish it. The German drug maker Bayer had undertaken a similar project and got nearly as desultory results (it was able to replicate 25 percent of the studies it reexamined). That study, published in a specialty journal in September 2011, hadn't sparked a lot of public discussion. Begley thought his study would gain more credibility if he recruited an academic scientist as a coauthor. Lee Ellis from the MD Anderson Cancer Center in Houston lent his name and analysis to the effort. He, too, had been outspoken about the need for more rigor in cancer research. When the journal *Nature* published their commentary in March 2012, people suddenly took notice. Begley and Ellis had put this issue squarely in front of their colleagues.

They were hardly treated as heroes. Robert Weinberg, a prominent cancer researcher at the Massachusetts Institute of Technology, told me, "To my mind that [paper] was a testimonial to the silliness of the people in industry—their naïveté and their lack of competence." When they spoke at conferences, Begley said, scientists would stand up and tell them "that we were doing the scientific community a disservice that would decrease research funding and so on." But he said the conversation was always different at the hotel bar, where scientists would quietly acknowledge that this was a corrosive issue for the field. "It was common knowledge; it just was unspoken. The shocking part was that we said it out loud."

The issue of reproducibility in biomedical science has been simmering for many years. As far back as the 1960s, scientists raised the alarm about well-known pitfalls—for instance, warning that human cells widely used in laboratory studies were often not at all what they purported to be. In 2005, John Ioannidis published a widely cited paper, titled "Why Most Published Research Findings Are False," that highlighted the considerable problems caused by flimsy study design and analysis. But with the papers from Bayer and then Begley, a problem that had been causing quiet consternation suddenly crossed a threshold. In a remarkably short time, the issue went from back to front burner.

Some people call it a "reproducibility crisis." At issue is not simply that scientists are wasting their time and our tax dollars; misleading results in laboratory research are actually slowing progress in the search for treatments and cures. This work is at the very heart of the advances in medicine. Basic research—using animals, cells, and the molecules of life such as DNA—reveals the underlying biology of health and disease. Much of this endeavor is called "preclinical research" with the hope that discoveries will lead to actual human studies (in the clinic). But if these preclinical discoveries are deeply flawed, scientists can spend years (not to mention untold millions of dollars) lost in dead ends. Those periodic promises that we're going to cure cancer or vanquish Alzheimer's rest on the belief that scientific discoveries are moving us in that direction. No doubt some of them are, but many published results are actually red herrings. And the shock from the Begley and Ellis and Bayer papers wasn't just that scientists make mistakes. These studies sent the message that errors like that are incredibly common.

At first blush, that seems implausible, which is perhaps one reason that it took so long for the idea to gain currency. After all, scientists on the whole are very smart people. Collectively they have a long record of success. Biomedical research is responsible for most of the pills in our medicine cabinets, not to mention Nobel Prize–winning insights about the very nature of our being. Many biomedical scientists are motivated to discover new secrets of life—and to make the world a better place for humanity. Some scientists studying disease have relatives or loved ones who have suffered from these maladies, and they want to find cures. Academics aren't generally in it for the money. There are more lucrative ways to make use of a PhD in these fields of science. Last but not least, scientists take pride in getting it right. Failure is an inevitable aspect of research—after all, scientists are groping around at the edges of knowledge—but avoidable mistakes are embarrassing and, worse, counterproductive.

The ecosystem in which academic scientists work has created conditions that actually set them up for failure. There's a constant scramble for research dollars. Promotions and tenure depend on their making splashy discoveries. There are big rewards for being first, even if the work ultimately fails the test of time. And there are few penalties for getting it wrong. In fact, given the scale of this problem, it's evident that many scientists don't even realize that they are making mistakes. Frequently scientists assume what they read in the literature is true and start research projects based on that assumption. Begley said one of the studies he couldn't reproduce has been cited more than 2,000 times by other researchers, who have been building on or at least referring to it, without actually validating the underlying result.

There's little funding and no glory involved in checking someone else's work. So errors often only become evident years later, when a popular idea that is poorly founded in fact is finally put to the test with a careful experiment and suddenly melts away. A false lead can fool whole fields into spending years of research and millions of dollars of research funding chasing after something that turns out not to be true.

Failures often surface when it's time to use an idea to develop a drug. That's why Glenn Begley's results were so jaw-dropping. That very high failure rate focused on studies that really mattered. Drug companies rely heavily on academic research for new insights into biology—and particularly for leads for new drugs to develop. If academia is pumping out dubious results, that means pharmaceutical companies will struggle to produce new drugs. Of course, Begley's test involved just fifty-three studies out of the millions in the scientific literature. And he chose those papers because they had surprising, potentially useful results. Perhaps a survey of more mundane studies would show a higher success rate—but, of course, those studies aren't likely to lead to big advances in medicine.

There has been no systematic attempt to measure the quality of biomedical science as a whole, but Leonard Freedman, who started a nonprofit called the Global Biological Standards

Institute, teamed up with two economists to put a dollar figure on the problem in the United States. Extrapolating results from the few small studies that have attempted to quantify it, they estimated that 20 percent of studies have untrustworthy designs; about 25 percent use dubious ingredients, such as contaminated cells or antibodies that aren't nearly as selective and accurate as scientists assume them to be; 8 percent involve poor lab technique; and 18 percent of the time, scientists mishandle their data analysis. In sum, Freedman figured that about half of all preclinical research isn't trustworthy. He went on to calculate that untrustworthy papers are produced at the cost of $28 billion a year. This eye-popping estimate has raised more than a few skeptical eyebrows—and Freedman is the first to admit that the figure is soft, representing "a reasonable starting point for further debate."

"To be clear, this does not imply that there was no return on that investment," Freedman and his colleagues wrote. A lot of what they define as "not reproducible" really means that scientists who pick up a scientific paper won't find enough information in it to run the experiment themselves. That's a problem, to be sure, but hardly a disaster. The bigger problem is that the errors and missteps that Freedman highlights are, as Begley found, exceptionally common. And while scientists readily acknowledge that failure is part of the fabric of science, they are less likely to recognize just how often preventable errors taint studies.

"I don't think anyone gets up in the morning and goes to work with the intention to do bad science or sloppy science," said Malcolm Macleod at the University of Edinburgh. He has been writing and thinking about this problem for more than a decade. He started off wondering why almost no treatment for stroke has succeeded (with the exception of the drug tPA, which dissolves blood clots but doesn't act on damaged nerve cells), despite many seemingly promising leads from animal studies. As he dug into this question, he came to a sobering conclusion. Unconscious bias among scientists arises every step of the way: in selecting the correct number of animals for a study, in deciding which results to include and which to simply toss aside, and in analyzing the final results. Each step of that process introduces considerable uncertainty. Macleod said that when you compound those sources of bias and error, only around 15 percent of published studies may be correct. In many cases, the reported effect may be real but considerably weaker than the study concludes.

Mostly these estimated failure rates are educated guesses. Only a few studies have tried to measure the magnitude of this problem directly. Scientists at the MD Anderson Cancer Center asked their colleagues whether they'd ever had trouble reproducing a study. Two-thirds of the senior investigators answered yes. Asked whether the differences were ever resolved, only about a third said they had been. "This finding is very alarming as scientific knowledge and advancement are based upon peer-reviewed publications, the cornerstone of access to 'presumed' knowledge," the authors wrote when they published the survey findings.

The American Society for Cell Biology (ASCB) surveyed its members in 2014 and found that 71 percent of those who responded had at some point been unable to replicate a published result. Again, 40 percent of the time, the conflict was never resolved. Two-thirds of the time, the scientists suspected that the original finding had been a false positive or had been tainted by "a lack of expertise or rigor." ASCB adds an important caveat: of the 8,000 members it surveyed, it heard back from 11 percent, so its numbers aren't convincing. That said, *Nature* surveyed more than 1,500 scientists in the spring of 2016 and saw very similar results: more than 70 percent of those scientists had tried and failed to reproduce an experiment, and about half of those who responded agreed that there's a "significant crisis" of reproducibility.

These concerns are not being ignored. From the director's office in Building 1 at the National Institutes of Health (NIH), Francis Collins and his chief deputy, Lawrence Tabak, declared in a 2014 *Nature* comment, "We share this concern" over reproducibility. In the long run, science is a self-correcting system, but, they warn, "in the shorter term—the checks and balances that once ensured scientific fidelity have been hobbled." Janet Woodcock, a senior official at the Food and Drug Administration (FDA), was even more blunt. "I think it's a totally chaotic enterprise." She told me drug companies like Amgen usually discover problems early on in the process and bear the brunt of weeding out the poorly done science. But "sometimes we [FDA regulators] have to use experiments that have been done in the academic world," for example, by university scientists who are working on a drug for a rare disease. "And we just encounter horrendous problems all the time." When potential drugs make it into the more rigorous pharmaceutical testing regimes, nine out of ten fail. Woodcock said that's because the underlying science isn't rigorous. "It's like nine out of ten airplanes we designed fell out of the sky. Or nine out of ten bridges we built failed to stand up." She rocked back and laughed at the very absurdity of the idea. And then she got serious. "We need rigorous science we can rely on."

Arturo Casadevall at the Johns Hopkins Bloomberg School of Public Health shares that sense of alarm. "Humanity is about to go through a couple of really rough centuries. There is no way around this," he said, looking out on a future with a burgeoning population stressed for food, water, and other basic resources. Over the previous few centuries, we have managed a steadily improving trajectory, despite astounding population growth. "The scientific revolution has allowed humanity to avoid a Malthusian crisis over and over again," he said. To get through the next couple of centuries, "we need to have a scientific enterprise that is working as best as it can. And I fundamentally think that it isn't."

We're already experiencing a slowdown in progress, especially in biomedicine. By Casadevall's reckoning, medical researchers made much more progress between 1950 and 1980 than they did in the following three decades. Consider the development of blood-pressure drugs, chemotherapy, organ transplants, and other transformative technologies. Those all appeared in the decades before 1980. His ninety-two-year-old mother is a walking testament to steadily improving health in the developed world. She is taking six drugs, five of which "were being used when I was a resident at Bellevue Hospital in the early 1980s." The one new medication? For heartburn. "You would think that with all we know today we should be doing a lot better. Why aren't we there?"

The rate of new-drug approval has been falling since the 1950s. In 2012, Jack Scannell and his colleagues coined the term "Eroom's law" to describe the steadily worsening state of drug development. "Eroom," they explained, is "Moore" spelled backward. Moore's law charts the exponential *progress* in the efficiency of computer chips; the pharmaceutical industry, however, is headed in the opposite direction. If you extrapolate the trend, starting in 1950, you'll find that drug development essentially comes to a halt in 2040. Beyond that point developing any drug becomes infinitely expensive. (That forecast is undoubtedly too pessimistic, but it makes a dramatic point.) The only notable uptick occurred around the mid-1990s, when researchers made some remarkable progress in developing drugs for HIV/AIDS. (The situation improved modestly in the years after Scannell and colleagues' analysis ended in 2010.) These researchers blame Eroom's law on a combination of economic, historical, and scientific trends. Scannell told me that a lack of rigor in biomedical research is an important underlying cause.

For Sally Curtin, it's personal. Crisis struck on February 5, 2010. She came downstairs in her eastern Maryland home to find her fifty-eight-year-old husband, Lester "Randy" Curtin, lying unconscious on the floor. She and an emergency crew fought through a blizzard to get him to the hospital. It took doctors four days to reach a diagnosis, and the news could hardly have been worse. Randy had a brain tumor, glioblastoma multiforme.

Both Sally and Randy worked at the National Center for Health Statistics (part of the Centers for Disease Control and Prevention). He was the guy colleagues went to when they were having trouble working through a statistical problem. When it came to his own odds, the doctors told them not to look at the survival numbers—but "we're numbers people," Sally Curtin told me. "The first thing we did was go look at the numbers." Half of patients with this diagnosis live less than fifteen months, and 95 percent are dead within five years.

"I had never heard the term glioblastoma. It seemed unreal to me that there was a cancer this lethal that they had not made progress on in fifty years," Sally told me. This cancer strikes about 12,000 Americans per year. (Senator Ted Kennedy was one of the most notable victims. Vice President Joe Biden also lost his son Beau to glioblastoma.) Even so, the Curtins hoped they could beat the odds. They signed Randy up for three separate clinical trials at the National Institutes of Health—experimental treatments that they hoped would keep the spreading tumors in check. None of them worked. In fact, in one brief period during the treatment, the tumors grew by 40 percent.

The worst part was that the disease was attacking the brain of a man with a powerful intellect. "His oldest daughter put it best. She said it's like telling someone who's afraid of the water that you are going to have death by drowning." With treatment options exhausted, Randy returned home to Huntingtown, Maryland, and registered for hospice. As the disease progressed, Sally said her husband had hallucinations. He would smash furniture, and once he pulled down the TV. "He really scared the kids," nine-year-old Daniel and eleven-year-old Kevin. "It wasn't like he was abusive or angry at us. He was just out of his mind" as the tumor grew. He hung on for seven months, increasingly agitated and in constant pain. Near the end, he asked Sally to overdose him with morphine, but she could not take his life. Eventually he slipped into a coma. At one point he had a seizure that jolted him out of it and was lucid enough to tell Sally, "I love you." That was the last thing he said to her. Five days later, he died, shortly after his sixtieth birthday.

Sally told me the story with strength and resolve. I wasn't surprised to learn that she had insisted on speaking at his memorial service, eight days after he died. She says she was able to keep her composure. Now in her early fifties, she's trying to figure out what the life of a widow is supposed to look like.

Glioblastoma provides a glimpse into the broader challenges facing biomedical science. Over the years, scientists have published more than 25,000 papers on the disease. The NIH spends about $300 million a year on brain cancer research. Scientists have made some headway in understanding the biology of this disease, but it simply hasn't translated into effective treatments—in part because the cells and animals studied in the lab are poor stand-ins for human beings. The failure rate may also reflect a lack of rigor in some studies testing experimental treatments in people.

At Arizona State University in Tempe, Anna Barker has been keeping score. Taped to the door of her light-filled conference room was a poster filled with print so fine I couldn't read the words from a short distance away. Barker told me it listed two hundred clinical trials that had been run on glioblastoma multiforme. Every single one was a failure. And the results of cancer studies overall aren't that much better. "Probably 65 to 80 percent of our trials in oncology fail," she told me. "Look at the money wasted. It's unbelievable." Barker has a personal passion for doing something about this. "Ultimately I lost my whole family to cancer, which is pretty amazing when you think about it." She was twelve years old when her grandmother died of pancreatic cancer, and "that was the reason I wanted to work in cancer research. But it never dawned on me as I grew up . . . that I would lose my sister and mother and father to cancer."

Before coming to Arizona State, Barker was deputy director of the National Cancer Institute. There she saw one big problem with cancer research: scientists were not approaching many studies with enough rigor. Each scientist had his or her own way of working, but those were not standardized or often repeatable. That's the culture of biomedical science today—researchers are individual entrepreneurs, each attacking a small piece of the problem with gusto. Barker says that unfortunately the quality of the work is all over the map—and there's typically no way to tell which studies you can believe and which you can't, especially when scientists try to add together results from different laboratories, each of which has used its own methods.

"Everyone says, 'It's not my problem,'" Barker told me. "But it has to be someone's problem. What about accountability? At the National Cancer Institute, we spent a lot of money. Our budget was $5 billion a year. That's not a trivial amount of investment. If a major percentage of our data is not reproducible, is the American taxpayer being well served?" Barker reads scientific journals with trepidation. "I have no clue whether to trust the data or not," she said. She has thought long and hard about how biomedical science got to this point and has some strong ideas about how to seize this moment to institute significant reforms. She's putting some of these ideas to the test by trying to revolutionize the treatment of glioblastoma brain tumors (more about that in Chapter 9).

Barker has good reasons to approach the scientific literature with caution. When an exciting scientific discovery is reported, scientists are quick to jump on the bandwagon, often without considering whether the original finding is in fact true. Here's a case in point. In 1999 and 2000, several scientists made a startling claim: they announced that bone marrow stem cells could spontaneously transform themselves into cells of the liver, brain, and other organs. "Transdifferentiation," as this was called, created instantaneous excitement because up until that point scientists had been harvesting from human embryos the stem cells they wanted for research. It seemed transdifferentiation could provide a much less fraught method. In short order, the scientific literature filled with dozens and eventually hundreds of papers backing up this rather remarkable finding. Some scientists even dropped whatever they'd been working on and started devoting their time to transdifferentiation.

The first splash of very cold water came from Amy Wagers, working in Irving Weissman's lab at Stanford University. First she irradiated mice to kill off their bone marrow cells and then injected them with a single bone marrow stem cell from another mouse—a cell that glowed green. That cell did indeed divide and create a variety of bone marrow cells (which is what these stem cells normally do, so no surprise). But it did not transdifferentiate into kidney, brain, liver, gut, muscle, or lung cells as previous experiments claimed it would. In a second experiment, she surgically connected the circulatory systems of several pairs of mice, with one in each pair containing the green glowing cells, and observed them for six to seven months. Wagers and her colleagues examined millions of cells in the recipient mice and again found no evidence of transdifferentiation. True, a few cells unexpectedly turned green (which other scientists had noticed in previous studies), but that was because cells were fusing with one another, not changing their fundamental identities. So much for the idea of transdifferentiation. In 2002, Wagers concluded with typical scientific understatement that transdifferentiation is "not a typical function" of normal stem cells found in the bone marrow.

Some researchers simply dismissed Wagers's discovery and kept on publishing their transdifferentiation results. It can be hard to give up on an idea, particularly if you've placed a heavy intellectual bet on it, investing time, reputation, and money. And scientists kept on reporting results that they thought were building their case. It was all a mirage. "Most of these studies turned out not to be reproducible," wrote Sean Morrison, a Howard Hughes Medical Institute investigator at the University of Texas Southwestern Medical Center. The mundane explanation for the initially exciting results actually had nothing to do with cells changing from one type to another. "This episode illustrated how the power of suggestion could cause many scientists to see things in their experiments that weren't really there and how it takes years for a field to self-correct," Morrison wrote in a scientific editorial, noting that scientists are sometimes too eager to rush forward "without ever rigorously testing the central ideas. Under these circumstances dogma can arise like a house of cards, all to come crumbling down later when somebody has the energy to do the careful experiments and the courage to publish the results."

Scientists concerned about this problem in biomedical research have come to call it the reproducibility crisis. But that term doesn't capture the true scope of what's happening. The scientific method, when properly used, doesn't simply apply to the conduct of a single experiment and ask whether it can be reproduced. The scientific method should also help researchers build a deeper understanding of biology and disease. It's not enough to know whether a particular discovery can be replicated using the exact same set of ingredients. Scientists want to find results that mean something more broadly. The overarching goal of biomedical research is to understand the basic processes that lead to disease so that medical science can intervene to ease human suffering and improve health.

That requires rigor in every individual experiment. But it also requires rigorous thought and insight putting those results into a broader context. And biomedical science is now suffering from a lack of rigor. Of course one way to measure rigor is to look at the first, fundamental step: testing whether individual studies can be reproduced. That's one reason Glenn Begley's paper struck such a nerve. And despite the hullabaloo it has caused, there has been a noticeable silence from the forty-six labs that produced the forty-seven papers Begley (and often they themselves) could not reproduce.

"None of the papers have since been retracted," Begley said. "No one has published follow-ups saying that the data was different. So I think probably they just felt that the first time was fine and something went wrong the second time. Many of these investigators had already moved on, which is typical in academia. They had moved on to whatever the next project was. So in most cases there was no real desire to set the record straight." In industry, exploring an idea for a new drug can quickly balloon into serious money, "so you can't just assume that you will be able to justify spending $100 million without replicating first."

When Begley undertook his project, Amgen signed secrecy agreements with the individual scientists, promising not to reveal their identities so as to spare them potential embarrassment. Other scientists have criticized Begley's work on those grounds—in point of fact, his report about reproducibility is not reproducible! Nobody can choose the same experiments and try to repeat Begley's work. He agrees with his critics that secrecy is a major shortcoming of his study. The only people who could disclose that information were the authors of the original studies, he said. "And they chose not to do so."

"I've invested my life in this area and this was just shocking," Begley says. "Since then, I've tried to do as much as I can. I hope that this will really change the way science is performed." After Begley's paper about irreproducible results made its splash, he heard from a postdoctoral researcher who pleaded with him to reveal the identities of those who did the flawed experiments. The young scientist worried that he was wasting his time working on a project based on one of them. Begley explained that the deal he'd cut with the researchers prevented him from exposing them, but the question did disturb him. His solution was to write a follow-up comment in *Nature* titled "Six Red Flags for Suspect Work." In it, he ran down the list of the six most common preventable failures he encountered. They're worth repeating here because they are very common failings found in biomedical research, and they explain a good deal of the reproducibility problem. Here are the questions that researchers should ask:

1. Were experiments performed blinded—that is, did scientists know, as they were doing the experiment, which cells or animals were the test group and which were the comparison group?

2. Were basic experiments repeated?

3. Were all the results presented? Sometimes researchers cherry-pick their best-looking results and ignore other attempts that failed, skewing their results.

4. Were there positive and negative controls? This means running parallel experiments as comparisons, one of which should succeed and the other of which should fail if the scientist's hypothesis is correct.

5. Did scientists make sure they were using valid ingredients?

6. Were statistical tests appropriate? Very often biomedical scientists choose the wrong methods to analyze their data, sometimes invalidating the entire study.

This list should be as familiar to scientists as the carpenter's dictum is to home builders: measure twice, cut once. Alas, the rules are often not applied. Training for a career in biomedical research is a haphazard process, with few formal courses. People learn from their mentors, for better or worse. In some fields, it's simply not tradition for scientists to follow these commonsense standards. For example, scientists studying mice in the lab may or may not believe it is important to assign their animals randomly to their study and control groups. And even when scientists follow these rules, they can still fail to generate reproducible results. Biomedical research is challenging even under the best circumstances.

Chapter Two

IT'S HARD EVEN ON THE GOOD DAYS

WHEN YOU STOP and think about it, experimental science can be a bizarre enterprise. In the classical view of research, scientists first come up with an exciting idea or make a provocative observation. Having done that, a good scientist will next actively look for evidence that his or her idea is *wrong*. Yet how disappointing it would be to discard your own seemingly wonderful idea. That's the first point at which human nature collides with the scientific process. In the words of the brilliant physicist Richard Feynman, "The first principle is that you must not fool yourself—and you are the easiest person to fool."

Scientists who can navigate those treacherous intellectual waters then face more daunting challenges: they must wade through the real-life human environment of academic science—funding, promotion, publication, and fame—which is full of perverse incentives that discourage them from probing deeply enough to find out whether their exciting ideas are actually wrong. Many of the problems in biomedical research today result when scientists often unwittingly stray from standard methods, so it's worth exploring how healthy science should work. Good methods not only test ideas; they help scientists avoid fooling themselves.

Careful science is a surprisingly young enterprise. Before the seventeenth century, natural philosophers, as scientists were then called, often relied on the word of authorities to sort out truth from fiction. For many hundreds of years, European intellectuals assumed that all knowledge already existed, and their job was simply to interpret the writings of the Greek philosopher Aristotle, who was looked upon as the ultimate authority. Around the time of Galileo (1554–1642), that edifice started to crack. Natural philosophers dared to conduct their own experiments to search for the truth. Some "facts" they examined seemed strange, like the widely held notion that you could heal a wound by putting ointment on the knife that caused it. A natural scientist realized he could test "strange" facts by trying to replicate them, as David Wootton explains in *The Invention of Science*.

A society formed in Florence after Galileo's death in 1642 took as its motto *provando e reprovando*, test and test again. Members announced their findings in the *Reports of the Society for Experiments*. Scientific publishing was born. British philosopher Francis Bacon had not long before formalized the scientific method: make a hypothesis, devise a test, gather data, analyze and rethink, and ultimately draw broader conclusions. That rubric worked reasonably well when scientists were exploring easily repeatable experiments in the realm of physics (for example, studies involving vacuum pumps and gases). But biology is a much tougher subject, since there are many variables and a great deal of natural variation. It's harder to see phenomena and harder to make sure personal biases don't creep in.

Here's a thought experiment to illustrate that point. Imagine you're locked into a windowless house in the woods. You can't sense day or night, and you can't feel the temperature outside. But you do have a reliable clock, and you can hear the birds. If you take meticulous notes, you can gradually discern a pattern showing which birds are singing and when. There's tremendous natural variability, of course. A late spring or short winter will throw off the pattern from one year to the next, but eventually you will be able to discern the seasons and determine the length of a full annual cycle. Of course it would be far better to fling open your door and enjoy actual nature in all its splendor, but that's not part of the bargain. Your conclusions must be inferences only. Likewise, a biomedical researcher can rarely witness directly the object of study. Life is mostly chemistry, and most of chemistry is invisible. Living cells also change depending on subtle variations in their environment, and those are hard to sort out as well. But with time, a picture gradually emerges from deductions based on indirect evidence.

"There is in fact no such thing as direct observation, for the most part," said Stanford University's Steven Goodman.

"Every scientific observation is filtered through an instrument of some sort." The tool may be an electron microscope, or it may be a clinical trial—in which the tool involves observing a group of human subjects—or it may be a statistical method. So the first question is, do you trust your tool to give accurate answers? "And if you don't believe that everything was properly done, you're not going to believe in the findings. So your ability to trust what you 'see' depends on your degree of trust in the instrument you're using." Scientists aren't simply evaluating plain facts, and their tools are rarely razor-sharp scalpels. Our observations of birdsong in the thought experiment will never be precise enough to determine that a year is 365.25 days. But bad ideas (like the hypothesis that a year lasts one hundred days) shouldn't survive the test of time.

"We might think of an experiment as a conversation with nature, where we ask a question and listen for an answer," Martin Schwartz at Yale wrote in an essay. This process is unavoidably personal because the scientist asks the question and then interprets the answer. When making the inevitable judgments involved in this process, Schwartz said, scientists would do well to remain passionately disinterested. "Buddhists call it non-attachment," he wrote. "We all have hopes, desires and ambitions. Non-attachment means acknowledging them, accepting them and then not inserting them into a process that at some level has nothing to do with you."

Sometimes this process produces observations that become readily accepted. For example, James Watson and

Francis Crick discovered that the DNA in our chromosomes is arranged as a double helix, with our genes encoded in the units that make up the rungs of the ladder. That observation is so clearly established as correct that it forms the basis for entire industries and fields of science. Nobody questions the structure of DNA, in part because it has proven so useful. But far more ideas linger in a world of twilight truth. Findings may be seen in one lab or several, but they don't easily become accepted as clear descriptions of nature, and they don't lead to useful insights for the treatment of disease. It can take many years for good ideas to rise to the top and bad ideas to drift to the bottom. And that limbo can be stretched out when experimental results from one lab clash with those of another. When done right, this process can yield deep insights into how biology works and lead to new ideas for maintaining health and treating disease. But it's a constant struggle for even the best scientists in the world to know whether they are fooling themselves.

The story of telomerase—a vital enzyme involved in aging and cancer—illustrates this point well. Carol Greider was a graduate student at the University of California, Berkeley, in the 1980s and working with her mentor, Elizabeth Blackburn, on a weird little single-celled pond critter called tetrahymena. The scientists were trying to understand how this microscopic protozoan, which looks like a hairy teardrop, manages to replenish the DNA at the ends of its chromosomes when it divides. When a cell divides, the chromosomes in its nucleus replicate as well. But with each cell division, the DNA at the chromosome ends gets a little shorter. That may seem like an obscure detail, but these chromosome tips are critical to life as we know it. And nobody understood how any organism could replenish its chromosome tips until Christmas Day 1984, when Greider discovered an enzyme in the tetrahymena that she thought could be doing the job.

"Rather than say, 'Look, let's find every piece of evidence we can to show that this is a new enzyme,' instead we did the opposite," Greider told me. "We said, 'How can we disprove our own hypothesis?'" She started to look for something else responsible for rebuilding the DNA at the chromosome tips. She was essentially trying to figure out whether she was fooling herself. When she tells this story to students, she reels off a long list of all the different steps she took to disprove her own hypothesis "because I would rather show that I'm wrong than have someone else show that I'm wrong." After an entire year of looking for flaws in their own work, she and Blackburn finally published the paper. Not only did her conclusion stand the test of time, her diligence led them to share the 2009 Nobel Prize for Physiology or Medicine.

The discovery of telomerase is a textbook case of science done right. But research on the enzyme also illustrates how difficult it can be for scientists to sort out competing ideas and move a field forward. Hundreds of scientists are now struggling to understand its role in biology and disease—and much of what they publish is contentious. A couple of years ago Greider was at a meeting dedicated to telomerase when someone declared, "Probably greater than 50 percent of what is published in the field just is not true." She agreed. Her former mentor Elizabeth Blackburn has published papers suggesting that meditation can make your telomeres grow and potentially extend your life span (Blackburn also founded a company to measure telomere length). Greider politely declined to talk about that unorthodox idea, but she spoke about other areas of telomere research that she considers questionable.

Indeed, she has spent a lot of time and money in her lab putting the findings of other scientists to the test. At one point she turned the tables on a scientist who had published another splashy paper about a unit of the telomerase enzyme called TERT. Steven Artandi and colleagues at Stanford University had engineered mice to produce an abundance of TERT and discovered that they grew a lot of extra hair. This apparently had nothing to do with telomeres, so the Stanford group proposed that TERT played another role in the cell, switching on and off genes unrelated to chromosome tips. Hair growth was just one example; other scientists suggested that TERT could flick on or off other genes.

Greider works with a physician colleague at the Johns Hopkins University School of Medicine, Mary Armanios, who treats people with diseases caused by defective telomerase. They realized that the ailments they were studying might be related to some of these surprising new results. So

Greider ran an experiment to put the Stanford results to the test. She started with a strain of mice with extralong telomeres. They could remain healthy for several generations without telomerase. She then deleted the TERT gene from these mice, not only disabling telomerase but also affecting all other functions related to TERT. Even so, these mice remained healthy for several generations. Based on that observation, she concluded that TERT doesn't play an essential role in gene regulation, as the Stanford scientists proposed. It is only essential as a part of telomerase.

So here's a case where two powerful tools of biology produce two different results. Neither tool is perfect—mice that produce far too much TERT aren't ideal, but neither are mice that are missing a gene. "It's not an issue of reproducibility; it's an issue of interpretation and understanding the mechanism," Steven Artandi told me. Does TERT actually turn genes on and off in normal tissue? "I'm not sure we proved it," Artandi acknowledged. "I think proving things takes time. So I think the jury is out." In Greider's view, the case is closed unless Artandi can come up with new data. This is all part of the normal, healthy process of science.

Like Carol Greider's lab, Tom Cech's at the University of Colorado, Boulder, spends a surprising amount of time doing studies that, in the end, just end up debunking results from other labs—setting the record straight rather than generating new discoveries. For example, his team discovered that a commercially available antibody used in dozens of experiments to flag the TERT protein in fact didn't work as advertised. The company that sold the antibody to research labs removed it from its catalog as a result, but many reports with false conclusions based on that errant ingredient remain in the scientific literature. These papers are intellectual land mines for scientists who aren't keeping fully abreast of the field—and given the avalanche of scientific publications, keeping up with any field is daunting.

Cech, a Howard Hughes Medical Institute (HHMI) investigator who shared the 1989 Nobel Prize for Chemistry, is utterly philosophical about this. To him, these stories simply illustrate that science has strong self-correcting mechanisms built in, and eventually the truth will emerge. But it's not always comfortable for a scientist to raise these issues. The people you criticize "might be reviewing your grants" and deciding whether you deserve funding, Cech told me. "They might make a decision about whether they'll give you a job offer. They might be reviewing some other papers of yours. So there's a tendency to be careful about being too negative about other people's work." Cech, as a heavyweight in this field, feels free to speak his mind. "I can be brave enough to pick up a microphone [at a conference] and say, 'We've tried to reproduce Bill's work at Harvard and we think it's completely wrong.' About six other people will then raise their hand and say, 'It's wrong. We wasted a year on it too.' People who've been quiet suddenly jump up and say, 'You're absolutely right.'"

Cech says his early career didn't involve chasing down so many false leads. His Nobel Prize–winning work involved experiments more in the realm of chemistry than biology: he showed that RNA could help catalyze biological reactions inside cells. "We were protected for a while against having to deal with this kind of stuff," Cech told me. Biology is much more fickle, however, and as his research has become less based on chemistry, "we now are seeing the dark underbelly of real biology research. But it is what it is." Discovering an error in someone else's work is "not a glorious day in the lab. That's a setback, and whether you're right or they're right, it's not fun, really." Some members of his lab told me that at least they learned something in the process of sorting out a problem. "They're being very generous" with that assessment, Cech said. "It's really sort of a pain."

One reason scientists have been slow to recognize the problems of rigor and reproducibility in biomedicine is that failure is an inevitable part of the process, so cautious researchers expect a lot of "discoveries" to be wrong. They see it every day in their own labs. Most experiments simply don't work. The equivalent of a .300 batting average at the lab bench would be phenomenal. A standard-bearer of this philosophy is Stuart Firestein, a Columbia University biologist and author of *Failure: Why Science Is So Successful*, which argues that science only advances when researchers try something, fail, and then learn from their failures.

According to his argument, if everything worked exactly as expected, scientists would simply be chasing their own tails, reinforcing existing ideas rather than finding new ones. Failure and ignorance propel science forward, Firestein argues, and many would agree on that point. He extolls the virtues of the self-correcting nature of science. Sure, lots of stuff that gets published turns out to be rubbish. "I don't see this as a problem but rather as a normal part of the process of experimental validation." He writes that if scientists took too much time to make sure their results were correct, publication would slow to a "virtual trickle." And even then, the results could still turn out to be flawed. It's the job of the entire community, not simply the scientist who makes a claim, to figure out what's right and what's wrong.

But in his exuberant defense of failure, Firestein treads into territory where his colleagues are less apt to follow. He thinks it's perfectly fine that of the fifty-three papers Glenn Begley at Amgen studied, only six could be reproduced. "This has been characterized as a 'dismal' success rate of only 11%. Is it dismal? Are we sure that the success of 11% of landmark papers isn't a bonanza? I wonder if Amgen, looking carefully through its own scientists' data, would find a success rate higher than 11%—or lower? And what did Amgen pay for those 6 brand new discoveries? Nothing. Not a penny." He continues, "This so-called dismal success rate has spawned a cottage industry of criticism that is highly charged and lewdly suggestive of science gone wrong, and of a scientific establishment that has developed some kind of rot in its core," Firestein writes (Firestein, *Failure* © Oxford University Press). Obviously he begs to differ.

There's no question that the scientific enterprise is usually self-correcting—in the long run. If time and money were no object, the truth would usually emerge from the cacophony of research papers that run the gamut from cringe-worthy to brilliant. Nobody's arguing that the constant flow of errors has brought science to a standstill. The concern, though, is that people with deadly diseases are watching their own lives slip away. And taxpayers, by way of the National Institutes of Health (NIH), don't have infinitely deep pockets. Funds poorly spent in one sloppy research lab could instead have been invested in rigorous studies elsewhere. Significantly, Firestein doesn't acknowledge that many errors in biomedical science—from the mislabeling of melanoma cells as breast cancer cells to the purchase of antibodies that don't perform as advertised—are easily preventable. I asked Begley about the failure rate at Amgen. "I expect that 90 percent of experiments do fail," he told me, not just at the company but everywhere. "Our business is about managing failure." But he said if experiments fail because the scientists were lazy, the work was sloppy, or the analysis was bad, "that's not the failure of an experiment—it's a failure of the experimenter." And that's what he says he saw in the dozens of studies that he could not reproduce.

Part of the everyday challenge of research is trying to avoid fooling oneself through bias. Inevitably it creeps into even the best scientific efforts. Bias is often impossible to avoid because it frequently involves pitfalls that scientists simply can't foresee. So it too is part of the fabric of scientific research. And there is a seemingly endless list of ways that a scientist can unconsciously inject bias. Surveying papers from biomedical science in 2010, David Chavalarias and John Ioannidis cataloged 235 forms of bias. Yes, 235 ways scientists can fool themselves, with sober names such as confounding, selection bias, recall bias, reporting bias, ascertainment bias, sex bias, cognitive bias, measurement bias, verification bias, publication bias, observer bias, and on and on. Biases are usually not deliberate or even a conscious choice.

One classic example is that, for many years, researchers favored using male mice because they found it more challenging to deal with the estrous cycles of females. Only many years later did they appreciate that they were deeply skewing some of their results by studying only males. Reporting bias is another common problem in biomedicine. Scientists are much more likely to report the results of an experiment that "worked" than one that failed, even though discovering the lack of an effect can be just as important as a positive finding. That tendency skews the biomedical literature, tilting it to create a publication bias that can grossly distort the purported effect of a drug.

Observer bias is another big problem. Scientists pursuing an exciting idea are more likely to see what they're looking for in their data, and that alone can completely skew the results. Medical researchers sometimes avoid this by running double-blind trials in which neither the experimental subjects nor the scientists themselves know who's taking a drug and who's taking a placebo. Blinding is also good practice in laboratory experiments, but it's not rigorously applied. Ken Yamada at the NIH says it's more likely to be used in obvious situations, such as when someone is peering through a microscope and making judgment calls about cell shapes. But many scientists who conduct animal experiments don't bother to blind those studies (or if they do, they don't bother to report this important fact in their publications). Yamada said since the issue of reproducibility bubbled up in the past few years, he has become much more aware of the need to blind his experiments. He now insists on it in his lab. "It is more work because you have to involve someone else to do the blinding or you need to have some randomization technique or something like that," he said. And tapping neutral lab members to do that work can distract them from their own projects. So it can be a nuisance, but "it's really worth it."

Sometimes bias arises because the effect scientists are trying to study can't be measured cleanly or quantified. Gregory Petsko at the Weill Cornell Medical College studies Alzheimer's disease. Scientists want to measure cognition in Alzheimer's studies, but Petsko says cognition "is not a *thing*. . . . I don't even know what that is, much less how you measure it. Those are serious reproducibility issues, and they stem not out of malfeasance . . . but rather out of our inability to understand something that we're trying to use."

Still, scientists need to find something to measure, particularly if they're trying to figure out whether a drug is having an effect. And often they're looking for a small one. A change of only 10 or 20 percent can have profound biological significance (a small change in body temperature, for instance, is likely to be disastrous). And it can be very difficult to see a small effect, particularly if you're measuring something highly variable. Consider your daily commute. If you drive, the time will vary simply because of traffic. If you wanted to know whether alternate routes were faster, it could take many trials before you would be confident that one was, on average, better.

Not only are the effects variable but the instruments for measuring them can be blunt. Muscular dystrophy is a disease affecting mostly boys that causes gradual loss of muscle control. Scientists studying it rely very heavily on a measure called the six-minute walk test to gauge how the disease is progressing and to test whether a potential drug is effective. Researchers put two pylons down in a hospital corridor and ask their young patients to circle them for six minutes. The researchers measure how far they travel during that time. Not surprisingly, children's performances can vary a lot. Researchers at Nationwide Children's Hospital in Columbus, Ohio, asked nine boys to perform the six-minute walk test. After they'd rested, four randomly selected boys were asked to do it again with a reminder to "walk quickly and safely." The other five were told that they'd get $50 if they could beat their previous time. The boys with the cash incentive improved a lot more than the other boys did. That may not sound surprising, but Eric Hoffman, a muscular dystrophy researcher at Children's National Medical Center in Washington, DC, was astounded by the amount of improvement a $50 cash incentive generated: the effect was bigger than scientists have seen for any of the drugs being developed for muscular dystrophy. He said the six-minute walk test is a terrible measuring stick (though he used saltier language he asked me not to repeat). "That's not something you want your whole drug development worldwide to depend on." But, for lack of anything better, the test is in fact the accepted standard (and boys are not tempted with $50 bills in the formal drug studies).

Even when bias is minimized, biomedical scientists always have to face the fact that nature is fickle. Ken Yamada at the NIH said he spent more than a year perfecting a technique to mass-produce a useful protein called fibronectin. He had the system humming along beautifully, "and then suddenly over the span of a few weeks my isolation techniques didn't work. I tried everything I could think of" but couldn't produce the protein at a useful rate anymore. "We are just dealing with biological systems," he said, "and sometimes these things happen." Yamada was never able to figure out why his process was so reproducible for a year and then suddenly just stopped working. He ultimately devised a new method. "It's kind of humbling," Yamada told me. The one saving grace was that he hadn't yet published his methods for mass-producing this widely used protein. "It would have been very embarrassing."

Every scientist can tell you a story about experiments gone awry for reasons they'd never expected. An undetectable change in water quality can disrupt experiments. Sometimes switching from one batch of nutrients to another can make a major difference. (Nutrients may be variable biological material as well—fetal bovine serum, for example, is commonly used to help cells grow in the laboratory.) One lab moved its precious genetically engineered mice to a new facility and watched with horror as they all died—apparently because the bedding for the animals was switched to a commonly used corn-based material. Olaf Andersen at the Weill Cornell Medical College told me he nearly lost a friendship over differing results published by his lab and that of a close colleague. Finally, after some bitter words, they decided to sit down and try to resolve the discrepancy. Sorting through the possibilities took months, but apparently the difference boiled down to this: Andersen cleaned his glassware with acid, while his colleague used detergent.

Much of the time, scientists never get to the bottom of the mystery. They find a work-around (or sometimes a whole new project) and move on. But Curt Hines couldn't simply deflect his problem. Hines was toiling away in a lab in Berkeley while his collaborators were doing complementary experiments in Boston. The experiments in this study required freshly gathered breast tissue (from women undergoing breast-reduction surgery). The scientists needed to isolate one of the many cell types found in healthy breast tissue as part of an experiment to study breast cancer. "We had tried shipping cells across the country, but the cells didn't really like that. You end up with a vial of dead cells," Hines told me. So the teams decided they would run the same experiment in parallel. But when they compared the results achieved by Hines's lab at Lawrence Berkeley National Laboratory and his collaborator's lab at the Dana-Farber Cancer Institute, they were crestfallen to discover that they didn't match. This was particularly troublesome because both labs had decades of experience in working with breast cells, so work in one lab should readily be reproduced in the other. "It didn't matter how we'd do it," Hines told me. "I'd get my profile, [Boston postdoc Ying Su] would get her profile." To get to the bottom of this, he tried to duplicate the Boston setup as closely as possible. In addition to the more traditional lab gear, "they were using a Cuisinart," Hines said. "I went down and got the exact same model of Cuisinart they were using. That didn't fix it."

Finally, after a year of struggling to solve this very basic problem, Hines's lab chief, Mina Bissell, said that the two scientists needed to get together in the same lab and work it out. They decided to meet at Hines's lab, in a modern glass-sheathed building near the Berkeley waterfront. After sitting side by side as they worked through what they thought was an identical protocol, they discovered why their results diverged. At one step in the process, Hines stirred the cells by putting them in a device that rocked them back and forth gently, while Ying used a more vigorous stirring system that involved a spinning bar. Both methods are used routinely in labs, so there was no reason to suspect that this mundane step would produce utterly different results, but it did. "It took a little bit of luck, a little bit of patience, but I'm also stubborn," Hines said. They published their tale in a scientific journal—a step that remarkably few labs bother to take, even when a mystery is probably plaguing labs elsewhere. Unfortunately, problems like this are so frequent in research that scientists may consider solutions too mundane to mention, and journals may not readily recognize them as scientific results worthy of publication.

Often, the problem scientists are trying to solve is not only extraordinarily challenging but also confusing. In cancer, for example, what kills people is usually not the initial tumor but the disease as it spreads—metastasizes—through the body. Your tax dollars have been hard at work exploring the key steps in this process. One critical step in metastasis occurs when a tumor cell manages to invade healthy tissue. Scientists believe if they can understand that process, they might be able to develop an anti-metastasis drug, which could revolutionize cancer treatment. Metastasis involves enzymes that eat through the structural framework of tissues, called the extracellular matrix. Hundreds of published scientific papers describe this process but report conflicting results. So in an attempt to sort out the story, the editors of the *Journal of Cell Biology* asked two NIH scientists,

Thomas Bugge and Daniel Madsen, to review the literature and answer one simple but important question: Do these tissue-dissolving enzymes come from the invading cancer cells, from the tissue that's being invaded, or both? Bugge thought this would be a simple exercise to bring some clarity to a muddled field. He was wrong. After realizing how many studies had been published on this subject, he and Madsen narrowed the field to just four enzymes in four major cancers: breast, colon, lung, and prostate. That still left them with nearly 250 studies to examine. And the results were all over the map. Some concluded that the enzymes came only from the tumor cells; some concluded that the enzymes came only from the surrounding tissue; some said they came from a combination of the two. One study used seven molecular probes to isolate the source of the enzymes and concluded that they were all coming from the surrounding tissue. Another used six probes and "found the exact opposite," Bugge told me. "They were trying to be very thorough. Those were very well-done studies. They just came to opposite conclusions."

Bugge came away from his study realizing that he had not only failed to arrive at a simple answer but discovered a deep problem affecting an important area of cancer research. "It was certainly never, at least for my part, meant to be a paper on reproducibility in science," Bugge said. "That's not the reason we started doing it. It was just halfway into the process we realized that is what it would have to be." I sat down with Bugge four months after his paper was published. He had just returned from a scientific meeting where many of the field's leading researchers got together to catch up. Bugge had long ago realized that the field was divided into camps: one believed the enzymes came from cancer cells, and the other didn't. Perhaps these competing mind-sets explain why nobody had previously tried to reconcile the conflicting studies. He figured that his finding would cause some soul-searching or at least reflection, "but there was no discussion. . . . Maybe people don't find this to be important," he said with a tinge of irony.

In the 1990s, pharmaceutical companies had spent millions of dollars trying to block matrix-dissolving enzymes, hoping to come up with a drug that would stop cancer from metastasizing. They all ended in failure. You'd think it would be worth understanding why, and that requires understanding this critical piece of biology. But scientists aren't rewarded for looking back—their careers depend on looking forward, toward the next big idea.

Given the complexity of nature and the inevitable limitations of the tools to study it, scientific disagreements can last years or even decades. (While arguably half of all studies in the telomerase field are wrong, for example, scientists undoubtedly disagree about which half to toss out.) "We only work on things that we don't understand," said Mark Davis, an HHMI investigator at Stanford. "And so that means by definition we're perpetually confused, and we're just trying to fight through the fog, and at least hold on to some things and illuminate some things that we hope are true, with the best tools that we have available." He said that's what makes his job so exciting. "It's different from making cars or something." The true art in science is figuring out which ideas are good and should rise to the top and which ideas to discard to move science forward. "The trouble is with human beings," he said. "Human beings have vested interests."

"If you're a professional in any area, your status in the field, especially in academia, is based on your perceived expertise in that area. So the last thing you want to hear is usually some punk kid undermining stuff you've worked on. So there's an inherent friction there." And that's not only true for an individual study; it can apply to entire fields. A transformative idea can be disruptive. Davis and his students had been working on one research paper for more than four years. Although deliberately vague about the unpublished ideas, he said his team had looked at some basic immunological findings observed in mice and often extrapolated to people. The new research shows that the mouse data don't apply to humans and are leading immunologists astray. But journals keep rejecting the paper, Davis said. "One comment came back: 'If this paper is published it will set the field back 10 or 20 years!' And I thought that was a really remarkable statement. But ultimately I interpreted it as a cry for help. If you're so insecure about your field that one paper could do so much damage, what have you got?

What are you so proud of here that could be swept away so easily?"

"Clearly the old guard will suffer with a paradigm change, and sometimes their whole career will go away," Davis told me. "That's a very real thing that can happen, especially in crappy fields where nothing will happen for years and years and suddenly something happens." Davis said that ten years ago human immunology fit his definition of a crappy field. There were a few talented scientists, but the field was intellectually adrift. "It was just dead. It was a wasteland."

The same was true for cancer immunotherapy, he said, despite the breathless promises about interferons (immune-system modulators), which ultimately delivered far less than hoped. There was a lot of serious effort, "but you also had all these charlatans. You had all these people who were saying, 'Well I'm too busy to do controls in my experiment because I'm curing cancer. Don't bother me with doing good science. I'm curing cancer.' But they weren't." Eventually, some scientists started cutting through the clutter. Researchers had identified parts of the immune system they call checkpoints and developed custom-built drugs, "checkpoint inhibitors" that help enlist the body's immune system to unmask and attack certain tumors.

The story of checkpoints emerged from a rigorous set of observations and was carefully confirmed in multiple laboratories. Drug companies were gradually able to take that knowledge to develop a powerful new class of anticancer drugs. Though still not effective for most people who take them, they do make a remarkable difference for a subset of patients. Davis said cancer immunology is no longer on his list of crappy fields.

This is a reminder of just how much is at stake. In retrospect we can see how much more rapidly these ideas would have advanced had it not been for the blind alleys and missteps along the way. To some extent, those meanderings are unavoidable—that's simply the nature of research on the frontier. But there are also many painful examples of delays, diversions, wasted time, and wasted money that scientists could have avoided had they been more careful along the way. And the failures begin right at the start—as scientists set out to design their experiments.

Chapter Three

A BUCKET OF COLD WATER

ONE OF THE most heartbreaking stories about abysmal experimental design involves amyotrophic lateral sclerosis (ALS), better known as Lou Gehrig's disease. The search for a treatment for this deadly degenerative disease is rife with studies so poorly designed that they offered nothing more than false hope for people essentially handed a death sentence along with their diagnosis. Tom Murphy was one of them.

Once an imposing figure, Murphy had played football and rugby in college. His six-foot-three frame and barrel chest gave him a solid presence. But his handshake wasn't the crushing grip you might expect. The first time we met, it was a gentle squeeze. When we met again a year later, we didn't shake hands at all. Murphy had lost his formerly impressive strength due to ALS.

People around the world donated more than $100 million to fight this deadly ailment during the Ice Bucket Challenge of 2014, but for most people its real-life consequences are an abstraction: something about the degeneration of nerves. For Murphy, a fifty-six-year-old father of three, ALS was a very concrete, slow march toward the day when his nerves could no longer direct his diaphragm to draw air into his lungs. (Physicist Stephen Hawking is the rare exception who has managed to survive for many years despite the diagnosis.)

Murphy, remarkably, was not bitter about this turn of events when he told his story. Nor was he resigned to fading away when he first noticed some unusual muscle twitches in the winter of 2010. He went to his doctor, who, after a brief examination, sent him to a neurologist. Murphy actually ended up seeing three different neurologists before he finally got the diagnosis.

"When the guy said, 'Sorry to tell you, but you have two to four years. Get your stuff together,' I thought, 'Really?' It was a real curveball. I would never have thought that in a million years." To prepare for what was likely to come, Murphy and his wife, Keri, sold the family home and bought a modern ranch-style house in Gainesville, Virginia, which Murphy could navigate without having to contend with stairs. He would eventually be getting around on wheels, once the muscle tone in his legs had faded. A giant TV graced the open and airy living room, where Murphy watched sports that he could no longer play himself.

But Murphy's doctors also offered at least a sliver of hope. "The first thing they told me is we have a drug trial; would you like to be in it? And of course I thought it sounded pretty good," Murphy said. People with ALS find their strength declines within a few years, and trials of potential drugs are only available to reasonably strong patients. So most only get one shot at an experimental treatment. In May 2011 he settled on the test of a drug called dexpramixole (or simply "dex"), becoming one of about nine hundred patients enrolled in a multi-million-dollar study. But when the drug company analyzed the data collected, the news was disappointing. Dex was not slowing the progression of symptoms in this group of patients. The trial was a bust.

Murphy was philosophical. There's no question the disease is a tough one to counteract. Almost everything scientists have tried for ALS has failed (other than one drug with very marginal benefit). So all scientists in the field have gone in knowing the likelihood of failure is high, but they didn't know exactly why until a nonprofit research center called the ALS Therapy Development Institute (ALS TDI) in Cambridge, Massachusetts, began investigating that question. Researchers there decided to look at the original studies to see what they could learn. They discovered that the original animal studies to test these drugs were deeply flawed. They all used far too few mice, and as a result they all came up with spurious results. Some experiments used as few as four mice in a test group. Sean Scott, then head of the institute, decided to rerun those tests, this time with a valid experimental design involving an adequate number of mice that were handled more appropriately. He discovered that none of those drugs showed any signs of promise in mice. Not one. His 2008 study shocked the field but also opened a path forward. ALS TDI would devote its efforts to doing this basic biology right.

Scott died of ALS in 2009 at the age of thirty-nine—the disease runs in his family. His successor, Steve Perrin, has carried on as Scott would have, insisting on rigorous animal studies as the institute's scientists search for anything to help people like Tom Murphy. And they're not simply taking the basic—and what should have been obvious—step of starting with enough mice in each experiment. Male and female mice develop the disease at somewhat different rates, so if scientists aren't careful about balancing the sexes in their experiments, they can get spurious results. Another problem is that the ALS trait in these genetically modified mice can change from one generation to the next. The scientists at ALS TDI look at the genetics of every single animal they use in an experiment to make sure that all are identical. "These variables are incredibly important," Perrin said. Other scientists had often overlooked those pitfalls.

To get robust results, Perrin's group uses thirty-two animals—and compares them to an untreated group of thirty-two more mice. Academic labs don't use large numbers of mice in their experiments in part because they cost a lot of money. Perrin said each one of these tests costs $112,000, and it takes nine months to get a result. If you're testing three dosages of a medication, each requires its own test. Perrin's institute has shown clearly that cutting corners here can lead to pointless and wasteful experiments. Even so, "we still get some pushback from the academic community that we can't afford to do an experiment like that," he said. It's so expensive that they choose to do the experiments poorly.

It's not fair to blame the scientists entirely for this failure. The National Institutes of Health (NIH) paid for much of this research, and funding was stretched so thin that scientists said they didn't get as much as they needed to do their studies. So they made difficult choices. As a result, funders, including the NIH, spent tens of millions of dollars on human trials using these drugs, without first making sure the scientific underpinnings were sound. ALS patients volunteered to test lithium, creatine, thalidomide, celecoxib, ceftriaxone, sodium phenylbutyrate, and the antibiotic minocycline. A clinical trial involving the last one alone, bankrolled by the NIH, cost $20 million. The results: fail, fail, fail, fail, fail, fail, fail. Science administrators had assumed that the academic scientists had all done the legwork carefully. They had not.

Of course a poorly designed study is simply a waste of time. Even so, it took years for officials at the NIH to realize the magnitude of this problem with ALS. One of the first people to pick up on this was Shai Silberberg at the National Institute of Neurological Disorders and Stroke (NINDS). He grew up in Israel and had trained as a biophysicist—a discipline that brings a high degree of precision to its studies, compared with the messier disciplines that involve live animals and people. That sensibility gave Silberberg a fresh perspective on the goings-on at the institute. The director asked him to serve on a committee that reviewed human tests involving neurological diseases, such as ALS, "and I couldn't believe what I was seeing," Silberberg told me. As he watched the review process, he discovered that the scientists spent all their time dwelling on questions about how to design the human trials—how many subjects, what the endpoints should be, making sure the analysis was framed correctly, and so on. All that's critical, of course. But Silberberg realized that nobody applied that same degree of care when it came to evaluating the animal studies upon which the human experiments were based. "I was in total shock," he said, when he realized that scientists basically skipped over those discussions. "There was almost no talk about whether the data to justify the [clinical trial] is solid or not." The assembled experts started out with the assumption that it was and were betting millions of dollars and the goodwill and lives of volunteers on the chance of that hope.

"I don't fault the reviewers, because clinical trials are so complex it's only natural that their focus was on the design, and if a disease is devastating, you kind of overlook this key part," Silberberg said. He felt "like the kid saying the emperor had no clothes." He convinced his boss, NINDS director Story Landis, that something was horribly wrong. Of course this was, to say the least, embarrassing. Human nature and institutional preservation both created incentives to hide or downplay problems. People on Capitol Hill looking for an excuse to slash domestic spending could cite this as an example of government waste. But Landis didn't back off. She started writing and talking publicly about the problem. Her boss, NIH director Francis Collins, was stunned when he heard about the pointless ALS trials that taxpayers had funded. "Humans were being put at risk based on that kind of data, and that took my breath away," he told me. It became readily apparent that the ALS story was not simply an isolated incident. Biomedical research had a problem, and Collins was, more than anyone, responsible for the enterprise. "Certainly there were people who didn't want to hear it. And I think there are still people who don't hear it," he said. But "as stewards of the public trust, we don't want to just sweep that under the rug. We want to face it square on and to be as transparent as possible and say, 'Okay, Houston, we have a problem here,' and we are all collectively going to have to face it."

Congress started to wake up to the issue. Republican senator Richard Shelby of Alabama raised it at a hearing on March 28, 2012. He brought up a December 2011 *Wall Street Journal* story based on the Bayer study of replication failures from that fall. "This is a great concern, Dr. Collins," Senator Shelby said at the hearing. "I don't want to ever discourage scientific inquiry, and I know you don't, or basic biomedical research. But I think we on this subcommittee, we need to know why so many published results in peer-reviewed publications are unable to be successfully reproduced. When the NIH requests $30 billion or more in taxpayer dollars for biomedical research—which I think is not enough—shouldn't reproducibility, replication of these studies, be a part of the foundation by which the research is judged? And how can NIH address this problem? Is that a concern to you?"

"It certainly is, Senator," Collins replied. The NIH director assured the Senate panel that he was on the case. Indeed, momentum for action was mounting fast. Only a few hours after that hearing ended, *Nature* published the even more devastating analysis by Glenn Begley and Lee Ellis. Collins assigned his chief deputy, Lawrence Tabak, to focus on the issue. The two acknowledged the issue plainly in a comment in *Nature* in January 2014, laying out a proposal to use NIH's leverage as the major funder of biomedical research to address the underlying problems. Those proposals gradually became new formal guidelines for grant applicants. As of January 2016, researchers must take some basic steps to avoid the most obvious pitfalls. When applying for a grant, they need a plan to show that the cells they are using are actually what they think they are (this is not a trivial issue, as we shall see). They need to show they've considered the sex of the animals they will use in their studies. They need to show that they've taken the time to find out whether the underlying science looks solid. And scientists must show in their applications that they will use "rigorous experimental design." Researchers are supposed to be held accountable for all this during the annual reviews of their grants. It's not clear how aggressively the various grant managers at NIH will enforce these new rules—officials historically have only canceled grants for egregious behavior, like fraud. So these steps are hardly cure-alls, but they are moves in the right direction.

It could take a long time for these new expectations to ripple through the culture of biomedical research. Grants written today could lead to research that won't be published for years. And of course there's likely to be resistance to any change with a whiff of more bureaucracy. Many academic scientists already spend more than half their time writing grant proposals, and because money is so tight, most of those don't get funded. People focused on treatments and cures aren't happy to wait around while scientists jump through hoops—even though research is not likely to succeed absent good experimental design.

Setting this system right requires changing the incentives. Steve Perrin at the ALS Therapy Development Institute said his operation is "the perfect paradigm for how to fix some of these problems." His institute focuses on treating a single disease and has hired a careful mix of people to chase that goal. It is also quite different from a university lab, where the work is done mostly by people in training: graduate students and postdoctoral fellows. Perrin uses staff scientists, not students. ALS TDI has another advantage over university labs, which must turn over a large percentage of their grant funding to their institutions: "We don't waste half of our investment on overhead," Perrin said. He doesn't even try to get grant funding from the NIH. Federal grant money is so tight these days that more than 80 percent of proposals get rejected, so it's not worth his while to have scientists devoting endless hours to writing grants. Instead, ALS TDI relies heavily on individual donors, especially people with a loved one with ALS or themselves stricken by the condition. (The wealthy board chairman who hired Perrin started funding the organization after he was diagnosed with ALS, and two members of the board had children with the disease.) "The biggest [fund-raising] challenge that we have in ALS is that our patients lose their battle with our disease very fast, which means our development team is constantly looking for new support," Perrin lamented.

Once they have the money, they know exactly how to put it to use. Given their expertise with the mouse model of ALS, they offer to reproduce the results from other laboratories, to validate—or most often to deflate—findings from academic and pharmaceutical labs. They have an aggressive program to develop their own drugs, based on tests carried out in their labs. The institute resides on the fourth floor of a modern lab and office building right across the street from the Massachusetts Institute of Technology (MIT). Next door are two other world-class institutions: the Broad Institute (which sequences genomes for ALS TDI) and MIT's Whitehead Institute. Perrin doesn't hesitate to farm out work that the many capable firms around Cambridge and the rest of the world can do more efficiently. Each Monday, the institute takes delivery of one hundred young genetically engineered mice, which are housed in a windowless expanse behind the light-filled and cheerful labs where nearly forty scientists work on artfully curved lab benches. Some spend their days doing experiments with the mice; others explore chemicals and biological compounds that are potential new drugs. This is the face of rigorous work, but rigor takes time and money—two commodities in perpetually short supply.

Federal rules are not acting alone in the push to improve experimental design. Disease advocacy organizations are playing an increasingly important role, with those for muscular dystrophy being a prime example. In 1986, Eric Hoffman and his academic mentor, Louis Kunkel, discovered the gene that goes awry in one common form of this disease, Duchenne muscular dystrophy. That launched Hoffman on his own respected academic career in the basic biology of this disease. In the 1990s, he was outraged when another medical researcher, Peter Law, enrolled desperate families in an experimental therapy for muscular dystrophy that Hoffman considered "snake oil," as he told me. (Hoffman's public protests were so rancorous that Law at one point sued him for defamation.) Law was injecting muscle-like cells into the young patients, despite overwhelming skepticism that this would work. "That was an incredible waste of resources and something that had no scientific basis at all," Hoffman told me. His moral outrage led him to push for some basic "standard operating procedures" for research on this disease: a standard that researchers worldwide would agree to and follow. To fund this, Hoffman turned to the US Department of Defense, which offers funding outside biomedicine's normal peer review process. He said he figured the NIH wouldn't give him a grant "because it is not hypothesis-driven, sexy research. It's developing rigor. There aren't many sources of funding for rigor."

His colleague Kanneboyina Nagaraju (known simply as "Raju") had also been vexed that sloppiness was common in muscular dystrophy research. He said almost all of it was done in academic labs "where the sample size is determined by the amount of money they had at the time." They'd call trials "pilot studies" and run them without proper controls, he said. With the money that Hoffman raised, Raju staged an international meeting, cosponsored by the European Commission, that developed consensus standards for the field.

Hoffman and Raju then built a laboratory at the Children's National Medical Center in Washington, DC, to run rigorous mouse tests that were part of this new standard. Scientists around the world realized that they could farm out these critical experiments to Raju rather than trying to perfect the technique in their own labs. This became such a popular service that Hoffman and Raju turned it into a small company (which moved to Halifax, Nova Scotia, in 2013). They have run tests on more than sixty potential drugs. Raju said fifty-five were found to be totally worthless, and the remaining five showed at least some promise. At one point, Raju said, a company gave him one drug to test and gave the same drug to a second lab in Italy running the same standard operating procedures. The company didn't mention that comparison until the experiment was completed. The results were not identical but close, he said, showing that the results are likely reproducible.

Frequently, small companies will plunk down a few hundred thousand dollars for these tests before investing even more in a potential drug. Raju said most companies accept the judgment from his studies. One French company that had only a single product in the works, closed shop after it got disappointing results, though others have ignored the clear warning signs, Raju told me. There's more at stake than just investors' dollars, Hoffman said. "It's the families. It's the patients. It's the physicians. It's the hospitals." Drug development is resource intensive "in terms of people's lives. They are the experiment. They become part of the drug development program." That resource should not be wasted. (Hoffman has since moved to Binghamton University.)

Debra Miller also has sent drugs to Raju for testing. She started Cure Duchenne after her son developed the disease in 2003. She and her husband decided to become venture philanthropists—investing in companies they believe in rather than simply funding academic research. In addition to hiring lawyers to do due diligence on these projects, her group hires top-notch researchers in the field. If something is going to fail, she told me, she wants it to fail fast. "So many small family foundations get seduced by the latest shiny object," she said, and they don't bother to insist on this kind of testing. "It all sounds like it's going to work." Miller knows better. Small biotech companies may nurse a dubious idea just long enough to sell the product to big pharma. Or companies may figure that it's easier to get a drug approved to treat a side effect of a rare disease like muscular dystrophy, when they really hope to use it for more common afflictions. Rare diseases like muscular dystrophy have a faster track through the Food and Drug Administration, so this is a potentially lucrative strategy. But parents don't want to spend their effort on drugs that may be only peripherally useful to their children.

When the advocacy group Parent Project Muscular Dystrophy was approached to fund research into a new drug, the group's leaders, John Porter and Sharon Hesterlee, insisted that the idea behind it first be tested using the standard operating procedures that Raju had developed. "Our mantra is we don't want clinical trials to fail for stupid reasons," Porter told me. Not following these basic procedures "is certainly one of the stupid reasons for clinical trials failing." More than two dozen companies are now working on drugs for muscular dystrophy, so there is plenty of promise, but also a need to figure out the best candidates. Porter, a former NIH official who dealt with these diseases, used a second layer of review if something passed initial muster:

"If a company comes to us with a project that they want to take to clinical trials, one of the first questions we ask is, has it been reviewed by TACT?"

TACT, which stands for the TREAT-NMD Advisory Committee for Therapeutics, was originally funded by the European Union but now runs with its own resources. It is a no-nonsense venue for reviewing potential drugs for neuromuscular diseases like muscular dystrophy. Twice a year, some of the world's experts in the field review submissions, ask tough questions, and render judgment—often harsh judgment. In the early days, many proposals came from academic researchers, but increasingly the reviews are done at the request of drug companies, which pay $5,000 or $10,000 to TACT to help defray expenses. The committee has reviewed dozens of proposals. Participants who elect to use this process get a confidential report that they can choose to share with potential funders and investors. A politely worded summary is posted on the public website, so at least the broad contours of a review are accessible even if a company decides not to share the details, which must be released in their entirety or not at all.

Getting biomedical research right means more than avoiding the more obvious pitfalls, like choosing the right number of animals, randomizing the experiments, blinding the observers so they don't fool themselves, and running proper comparison groups. It's also critical to think about whether the underlying assumptions are correct. The story of ALS offers a sobering example. The mice used in these studies have a specific mutation, called SOD-1. That trait shortens their lives and gives them some symptoms suggestive of the condition, but they do not in fact develop true ALS. Scientists developed this mouse model after discovering the SOD-1 mutation in some people who have an inherited form of ALS. But only 2 percent of people with ALS carry this mutation, so it's hardly the whole molecular story behind the disease. And that means it's not clear exactly what value comes from all the painstaking work with these mice by ALS TDI and many other labs.

The SOD-1 mice are used because scientists have had nothing better to model this disease. They are painfully aware of that. In fact, they've developed new strains of mice with mutations seen more commonly in ALS patients to get around the shortcomings of SOD-1 mice. But those new animals have shortcomings of their own: they don't die prematurely, which means it's much harder to study them in experiments where the endpoint is early death.

This is another reminder that no tool in biomedical research is perfect, so scientists always have to make do. And they may not take the time to question the assumptions on which their field rests. Scientists most often start a research project by building on what their mentors or peers have done before. An entire field may rely on a particular animal model, even though scientists often have no idea whether it's a valid surrogate for human disease. It's quite common to cure a disease in a mouse model, only to discover that it's irrelevant for treating human disease. And that, for scientists trying to conduct rigorous scientific research, makes mice a big, hairy problem.

Chapter Four

MISLED BY MICE

MISLEADING ANIMAL STUDIES have led to billions of dollars' worth of wasted effort and dead ends in the search for drugs. Failures in animal studies have also had deadly consequences. In 1993, researchers at the National Institutes of Health (NIH) wanted to test a potential drug for hepatitis B, a liver infection that affects hundreds of thousands of Americans. The compound, called fialuridine (FIAU), looked promising. It was similar in action to some of the drugs that had been developed to fight HIV, and it had passed animal tests with flying colors. Researchers gave it to mice, which tolerated it well. They then gave it to rats, which also had no trouble with the drug. Tests in monkeys also suggested the drug would be safe. A short-term study in humans was largely encouraging, so researchers enlisted fifteen volunteers to take the drug for several months.

At first, the volunteers managed the relatively mild side effects. But after a couple of months, one fell ill. It was liver failure. Since a toxic drug reaction can often cause liver failure, the researchers immediately told the remaining fourteen patients to stop taking the FIAU. But it was too late. In the following weeks, six more volunteers developed serious liver damage. Five people eventually died, and the Institute of Medicine, which reviewed this debacle, said the two other patients probably would have died as well if they hadn't received liver transplants.

At root, the problem is that lab animals aren't just small, furry humans. Researchers have been well aware of the underlying shortcomings of animal research for decades. To cite just one dramatic example, researchers working to develop a cholesterol-lowering drug thought they had hit on just the right compound after many years of laboratory research. But when they gave it to rats, it was an utter failure. One pharmaceutical company gave up on it entirely. But a Japanese researcher persevered despite the apparently show-stopping animal results. Arika Endo asked a colleague who was using chickens for experiments to test it, and it turned out the compound worked in the birds. That success marked the birth of statins, drugs used by millions of people today.

Stories like this teach an obvious lesson, especially for the most common laboratory animal, the mouse. "Nobody knows how well a mouse predicts a human," said Thomas Hartung at Johns Hopkins University. In fact a test on mice doesn't even predict how a drug will work in another rodent. For instance, certain drug-toxicity tests run separately on rats and mice only reach the same conclusion about 60 percent of the time. And if mice do a so-so job of predicting what will work in a rat, Hartung said, we should be very humble about what they tell us about human beings.

This is about more than drugs. Hartung said roughly half of the chemicals that show up as potential cancer-causing agents in mouse experiments are probably not human health hazards. Coffee is one example. Researchers have tested thirty-one compounds isolated from coffee, and of those twenty-three flunked the safety test. "You wouldn't be able to add coffee to food if it were synthesized" and put through a battery of safety tests, Hartung said. "Aspirin also would fail almost all animal tests these days."

Likewise, exciting results in animals often don't apply to human disease. A wave of enthusiasm followed the discovery of a weight-controlling hormone, leptin, in mice. Mutations in this gene caused obesity in mice, and giving those mice leptin slimmed them down. So scientists eagerly looked for this same effect in people, but leptin supplements rarely helped. There's no miracle pill to treat people who are overweight.

Biomedical scientists understand these shortcomings but tend to gloss over them. "You need to publish, and you can't write, 'I used a terrible model.' You have to do grant applications and you want awards, so you only show the good sides," Hartung said. "You're penalized if you do it differently, if you're honest about weaknesses." Experiments with mice and rats are everywhere in biomedical research. Nobody keeps a tally of the numbers used, but often-quoted estimates put the figure in the United States alone at well over 10 million animals a year, the vast majority being mice. One reason everybody uses mice: everybody else uses mice. They are "model organisms" used in basic biology studies as well as for safety research on experimental drugs. Scientists have developed hundreds of inbred strains of mice by breeding siblings with each other to perpetuate a genetically pure line of animals. Other strains are genetically engineered, allowing scientists to add or subtract specific traits. Entire industries have grown up around the breeding, shipping, housing, feeding, and care of mice.

Malcolm Macleod, a neurologist at the University of Edinburgh, worries about biomedicine's dubious reliance on mice. He has spent much of his career trying to find ways to reduce the brain damage caused by strokes. In hundreds of animal experiments, mostly using mice, drugs have shown promise for treating stroke. Billions of dollars have gone into this research, and yet not one single drug acting on brain cells has been shown to be helpful when tested in people (the drug tPA is effective by breaking down clots and helping restore blood flow, but it doesn't act on nerve cells damaged by a stroke). Neurologists started calling this long string of failures a "nuclear winter" for stroke research.

"My reading of the animal data for stroke is that it's not possible to say if they're good or bad models," Macleod told me. It could be that experimentally induced strokes in mice are so radically different from a human stroke that there are no real lessons to learn. Or it could be that the experiments have been carried out so loosely that they have led the entire field astray.

One example in particular stands out. AstraZeneca had high hopes for a drug called NXY-059. Researchers ran twenty-six experiments involving 585 animals (mostly rats), and in this case the compound appeared to protect the animals' brains from strokes induced by researchers. On the strength of those results, researchers tested the drug in an enormous and ambitious study involving more than 1,700 people. The results, which showed a modest reduction in disability, were sufficiently encouraging that they tested another 3,300 people who had just suffered strokes. It was a dramatic failure.

Macleod dissected that study and found many shortcomings. Failures like this have turned him into something of a crusader. At one point he decided to find out how widespread preventable biases had become in animal research. In 2015 he and his colleagues sampled animal research papers authored by researchers at the top universities in the United Kingdom. Only 17 percent were blinded. Likewise, scientists often failed to assign their animals randomly to drug versus control groups and to state whether they had conflicts of interest. Almost none explained how they settled on the number of animals they used in their studies. "It is sobering that of over 1,000 publications from leading UK institutions, over two-thirds did not report even one of four items considered critical to reducing the risk of bias," Macleod and colleagues wrote, "and only one publication [out of 1,000] reported all four measures."

Those problems are not hard to fix, so there's room for rapid reform. But those improvements won't help in the many instances when animals are poor stand-ins for human disease. In the case of stroke, Macleod noted that adolescent, male animals often used in the studies may not be a good substitute for an elderly human being having a stroke. And drug doses for animals may differ dramatically than those for people. He cautioned that the extent of brain damage in an animal may not be a good surrogate for human disability or death. Researchers, he admonished, need to think about these deeper issues, not just about whether they can run the same experiment and get the same results.

Stroke research is far from the only area stymied by dubious animal models. Pain studies, which also rely heavily on mice, have reached similar dead ends. A decade ago, pharmaceutical companies got very excited about a new class of pain medications called NK-1 antagonists. Experiments involved measuring a drug's effect by tying off nerve bundles in the animals to make them hypersensitive to pain. Using that method, the drugs seemed quite effective at reducing pain in the mice. That triggered a race among drug companies to come up with painkillers to block this pathway. Companies spent many millions of dollars trying to turn this exciting mouse research into a human drug. "People took this to the clinic and it didn't do anything.

Nothing at all," Barbara Slusher at Johns Hopkins told me. The mouse test was completely misleading when it came to judging real pain in people. One of the more spectacular drug-development failures in recent decades, it led researchers to realize they needed a whole new strategy to look for effective painkillers.

Slusher said it also soured the pharmaceutical industry on putting too much faith in animal studies. "It used to be that if your drug worked in an animal model, okay, we're cool." Drug companies would sign a deal with the researchers and move the potential drug forward in the development pipeline. Now pharma cares less about animal studies in pain research and instead wants human tests to validate an idea before making an investment. "So this is a change in the mind-set of how new targets are chosen." It's a sensible step, but of course it puts far more onus on academic researchers to do rigorous work.

Yet another great disappointment involves the treatment of deadly inflammation, related to trauma, burns, and other serious injuries. Sepsis and related conditions strike more than a million Americans each year and claim more than 200,000 lives. Geneticist Ron Davis at Stanford was curious to understand why there had been so little progress in finding drugs to stop that often fatal cascade of events. Heading up a small army of colleagues, he set out to test whether the commonly used mouse model for inflammation was a good match for the human disease. About 150 antisepsis drugs have been developed using mice, but not one has been helpful in people. The researchers identified about 5,000 genes activated or deactivated by inflammation in humans who had suffered trauma, burns, or blood infections. When they then looked at the analogous genes in one common strain of mice, they found essentially no correlation between the genes in mice with induced inflammation compared with humans. The biology of inflammation seemed to be dramatically different between the two species.

"That was a bit of a shocking result," he told me. It suggested that decades of inflammation research using mice was misguided and that scientists who continue to use mice for this research could be wasting their time. It was not a message the sepsis field wanted to hear. "It's amazing how much pushback we got from that result, including from within our own trauma center," Davis said. It was seen not simply as a provocative observation about inflammation but a frontal attack on studies involving mice. "There's a worry that mice will be invalidated," he said.

It's not uncommon for scientists to resist disruptive ideas, but "the power of science is, the truth will eventually come through," Davis said. "It's just a matter of how long that will take. You can suppress things, but you can never win." Some researchers have pushed back against Davis's findings, while other observers suggest there's even more at play. David Masopust at the University of Minnesota exposed laboratory mice—which are usually kept in sterile surroundings—to wild mice carrying germs typical of the species and discovered that their immune systems were dramatically different. So the differences between laboratory mice and people may be both genetic and environmental.

Despite all these pitfalls, scientists aren't interested in abandoning research with mice. Buried under Stanford's central campus is a labyrinth of tightly controlled rooms, stacked floor to ceiling with mouse cages. Visitors must gown up and slip booties over their shoes—not for their own health but to protect the untold thousands of mice that live there, under a vast, bucolic landscape punctuated with palm trees. Joseph Garner, who studies animal and human behavior, walked down a corridor and poked his head in one rodent-filled room after the next. Mice huddled together in clear plastic cages connected to an elaborate ventilation system to prevent germs from circulating. Even so, a musty smell typical of animal labs permeated the rooms, somehow sweet and stale at the same time, like home-brewed beer. Labels and barcodes keep this massive enterprise organized.

Garner said that mice have great potential for biological studies, but at the moment, he believes, researchers are going about it all wrong. For the past several decades, they have pursued a common strategy in animal studies: eliminate as many variables as you can, so you can more clearly see an effect when it's real. It sounds quite sensible, but Garner believes it has backfired in mouse research. To illustrate this point, he pointed to two cages of genetically identical mice. One cage was at the top of the rack near the ceiling, the other near the floor. Garner said cage position is enough of a difference to affect the outcome of an experiment. Mice are leery of bright lights and open spaces, but here they live in those conditions all the time. "As you move from the bottom of the rack to the top of the rack, the animals are more anxious, more stressed-out, and more immune suppressed," he said.

Garner was part of an experiment involving six different mouse labs in Europe to see whether behavioral tests with genetically identical mice would vary depending on the location. The mice were all exactly the same age and all female. Even so, these "identical" tests produced widely different results, depending on whether they were conducted in Giessen, Muenster, Zurich, Mannheim, Munich, or Utrecht. The scientists tried to catalog all possible differences: mouse handlers in Zurich didn't wear gloves, for example, and the lab in Utrecht had the radio on in the background. Bedding, food, and lighting also varied. Scientists have only recently come to realize that the sex of the person who handles the mice can also make a dramatic difference. "Mice are so afraid of males that it actually induces analgesia," a pain-numbing reaction that screws up all sorts of studies, Garner said. Even a man's sweaty T-shirt in the same room can trigger this response.

Behavioral tests are used extensively in research with mice (after all, rodents can't tell handlers how an experimental drug is affecting them), so it was sobering to realize how much those results vary from lab to lab. But here's the hopeful twist in this experiment: when the researchers relaxed some of their strict requirements and tested a more heterogeneous group of mice, they paradoxically got more consistent results. Garner is trying to convince his colleagues that it's much better to embrace variation than to tie yourself in knots trying to eliminate it.

"Imagine that I was testing a new drug to help control nausea in pregnancy, and I suggested to the [Food and Drug Administration (FDA)] that I tested it purely in thirty-five-year-old white women all in one small town in Wisconsin with identical husbands, identical homes, identical diets which I formulate, identical thermostats that I've set, and identical IQs. And incidentally they all have the same grandfather." That would instantly be recognized as a terrible experiment, "but that's exactly how we do mouse work. And fundamentally that's why I think we have this enormous failure rate."

Garner goes even further in his thinking, arguing that studies should consider mice not simply as physiological machines but as organisms with social interactions and responses to their environment that can significantly affect their health and strongly affect the experiment results. Scientists have lost sight of that. "I fundamentally believe that animals are good models of human disease," Garner said. "I just don't think the way we're doing the research right now is."

Malcolm Macleod has offered a suggestion that would address some of the issues Garner raises: when a drug looks promising in mice, scale up the mouse experiments before trying it in people. "I simply don't understand the logic that says I can take a drug to clinical trial on the basis of information from 500 animals, but I'm going to need 5,000 human animals to tell me whether it will work or not. That simply doesn't compute." Researchers have occasionally run large mouse experiments at multiple research centers, just as many human clinical trials are conducted at several medical centers. The challenge is funding. Someone else can propose the same study involving a lot fewer animals, and that looks like a bargain. "Actually, the guy promising to do it for a third of the price isn't going to do it properly, but it's hard to get that across," Macleod said.

There's an intellectual tug-of-war in biomedicine now about how best to increase the rigor of these studies. Should we improve animal studies? Or would we be better off simply trying to find replacements? Neuroscience professor Gregory Petsko at Weill Cornell Medical College is in the latter camp. He has spent his career studying neurological diseases, including ALS and Alzheimer's. "The animal models are a disaster," he said. "I worry not just that they might be wrong. 'Wrong' animal models you can work with. If you know why it's wrong, you can use the good parts of the model, and you don't take any information from the parts that are bad. But what if the neurodegenerative disease models are not wrong but irrelevant? Irrelevant is much worse than wrong. Because irrelevance sends you in the wrong direction. And I think the animal models for nearly all the neurological disorders are in fact irrelevant. And that scares the shit out of me, if you pardon the expression."

Researchers are so dependent on mouse experiments in neurological studies that if a potential compound doesn't work in the animals they won't try the much more expensive and time-consuming experiments on people. "That's what keeps me up at night—the possibility that I might have something that works but I would never be able to prove that it works," Petsko said. When you consider that many experiments have weaknesses above and beyond simply using mice, "it's not beyond the realm of possibility that a viable treatment already has been tested and failed because the trial design was bad. . . . I wouldn't rule it out absolutely." These aren't abstract questions. "You've got to do this right."

Petsko's hopes lie in a burgeoning technology called induced pluripotent stem (iPS) cells. Those cells, taken from a patient with a particular disease (say, Alzheimer's), can be induced in the lab to become nerve cells, genetically identical to the patient's but now growing in the lab. Petsko would like to use those cells to study disease biology and to reserve animal tests for safety and for insights into how a potential drug interacts with an entire individual. These iPS cells are rapidly gaining popularity. But like everything in biomedical research, this tool has its pluses and minuses. The downside is, once triggered to become a nerve cell, they continue to divide and proliferate—exactly the opposite of what happens to a nerve cell in the brain. So it's not clear how faithfully they mimic reality.

Thomas Hartung has been experimenting for a while with brain cells that proliferate in the lab but also morph into different brain cell types, forming round clumps of cells, called organoids. The cells in his lab come from patients with autism and Down syndrome. The clumps apparently can't think—though they do generate electrical signals, just like brain cells, and organize themselves in a manner reminiscent of how they are juxtaposed in the brain. They also use the chemical signals that underlie brain function. "If we create conditions in cell culture which are mimicking the organism, we are more likely to get relevant results," he told me. "You can do personalized toxicology with these cells. If I took your cells, I could tell you are more sensitive than another person to certain drugs, for example." These are early days for this technology, but there's a rapidly growing industry around cultivating disembodied blobs of cells in the lab. The Defense Advanced Research Projects Agency, which funds far-out ideas, has poured money into this line of research at multiple labs. So has the NIH. And Hartung has private money to work on the problem as well.

A company on the Boston waterfront is now mass-producing a related technology for use by pharmaceutical companies and academics alike. Emulate is a spin-off of the Wyss Institute at Harvard. The company has taken over industrial space in an old building that overlooks a working dry dock. Imposing concrete pillars with flared tops punctuate the floor space. The army used this building during World War II to assemble tanks. When Emulate moved in its high-tech manufacturing equipment, the landlord didn't bother to ask how much anything weighed. The freight elevators were built to hold a battle tank, and the concrete floors are nineteen inches thick.

Instead of growing spherical clumps of cells, Emulate builds ersatz organs using clear plastic chips that fit easily in the palm of a hand. Engineers swathed in white coveralls position themselves at laser cutters and 3-D printers that are designed to churn out prototype chips. Geraldine Hamilton, the president and chief scientific officer, excitedly showed me one chip with a flexible membrane. "In some organs, stretching is one of the most relevant mechanical forces," she explained, so that has been recreated as closely as possible for cells like those in the lungs, where growth and development depend on that kind of movement. Liver cells don't care about stretching, but they do react to how fluids flow past. So liver chips instead include an intricately arranged micro-irrigation system to control nutrients as they flow in and waste products as they flow out.

Hamilton told me they'd managed to keep a miniaturized liver alive and healthy for more than a month. It would be of absolutely no use to someone in need of a transplant, but these chips can be used to mimic liver biology and to test drugs. People have been trying experiments like this in the lab for many years, often with disappointing results. "When you put a drug into a dish, it just sits there on top of the cells. That's not the way we get exposed to drugs." The flow-through system is much more realistic.

These systems can be remarkably lifelike. The scientists introduce a single layer of cells onto a chip, but over the course of a week, elaborate structures grow. In the gut-on-a-chip, the natural folds found in the intestine spontaneously appear. Different types of cells form layers, much as they do in the human body, and closely resemble normal tissues under the microscope. Hamilton popped open a laptop and showed me a movie of the lung-on-a-chip in action. Tiny hairs, called cilia, waved like a kelp forest or a field of wheat. "Not only are the cilia beating just like you'd expect in your lung . . . you can see them clearing the particles in a directional manner," she said with a touch of pride and awe. She showed me a video of the immune system cells in action within a lung-on-a-chip, fighting back an infection. "We can actually see a single white blood cell. It comes in, it sticks, it wiggles itself through the cell membranes . . . and you can see it coming out on the other side and engulfing the bacteria."

"I've seen this ten thousand times, and it doesn't ever get old," said Chris Hinojosa, Emulate's principal microfluidics engineer. Daniel Levner, the company's chief technology officer, agreed: "This video was the aha moment for this technology." As just one hint of how it could be useful, Hamilton's group treated these lungs-on-a-chip with steroids in an effort to understand why people with chronic obstructive pulmonary disease resist this type of drug. "We were able to mimic that and look for potential therapeutics," Hamilton told me.

She hopes these techniques will prove so effective that the FDA will someday accept them in place of animal studies. That will not happen anytime soon, but Hamilton told me pharmaceutical companies are already starting to use chips in place of mice for experiments that won't need the FDA's stamp of approval.

Emulate offers lessons about the broader rigor and reproducibility issues. It's not just about devising a system that has the potential to sidestep some of the pitfalls of animal research. Levner said company scientists have to meet a higher standard than their counterparts in academia. "The tongue-in-cheek joke about the postdoc is you have to repeat [an experiment] three times—which is all you need for statistical significance for your *Nature* paper and your faculty position—and you're done. Doing it three times is very difficult, but it's also difficult to make three times into thirty times into three hundred times into three thousand times." Academics don't have that motivation, but this company needs to meet that mark to be successful. "You have your reliability [and] reproducibility," added Hinojosa, "and you know that when you see something, it's a real thing."

It's easy to be lured by the idea that new technology will swoop in and resolve the shortcomings in animal research.

That has sparked perpetual hope, with a history that includes development of specialized mouse strains and increasingly sophisticated strategies such as growing human organs (and tumors) inside mice. Indeed, technology is racing ahead throughout biomedical research. DNA sequencing technology has generated avalanches of data (recalling as well that avalanches are dangerous). Microscopy has opened our eyes to a fascinating world within living tissues. These tools provide deep insights into biology, but there is a startling disconnect between technological progress on the one hand and medical progress on the other. Technology is racing ahead at breathtaking speed, but medical progress is not.

Jack Scannell has been thinking about that a lot. His career has taken him from the world of pharmaceuticals to Wall Street and now to the University of Edinburgh. Scannell suspects rapid technological evolution is actually part of the problem. The more scientists come to understand the basics of biology, the more seduced they become by the idea that they can find cures by unearthing the fundamental mechanism of disease. This has deep intellectual appeal and could be true if we were approaching a complete understanding of how biological systems operate. But most of the time we only get a faint glimmer. Scientists may discover a single enzyme system involved in cancer and then hunt down molecules that can block that enzyme to treat disease. That raises hopes.

Every once in a while, something actually pans out—the anticancer drug Gleevec was a spectacular success built on this very idea. But most of the time those strategies fail. "So you got some spectacular successes which people remember, and a whole bunch of failures which people forget," Scannell told me. Instead, people look at the successes as signs that the molecule-by-molecule approach is working, when he argues it's almost always failing. "There is an almost automatic assumption that you need to rummage around in molecules to really understand things." But that actually accounts for a tiny share of medical advances.

Why do most ideas fail? Scientists tend to chalk it up to bad luck. But Scannell argues that evolution has created so many redundant systems that targeting a single pathway in a complex network will rarely work. Diet drugs are a good example. "We have evolved seventeen different biological mechanisms to avoid starving to death [figuratively speaking]. Drugging one of those mechanisms isn't going to do anything!" The fact that there are many pathways to cancer also explains why chemotherapy drugs often work for a while and then fail. The tumor evolves a work-around.

Back in the decades when drug development was progressing rapidly, doctors weren't trying to create new drugs based on a deep understanding of biology. They just experimented on people—not mice—to see what worked. "I wouldn't necessarily seek to defend the historic approach," Scannell said. "I think today people would be horrified if they knew how drug discovery really worked in the fifties and sixties. But I also think it is a historical fact that it was an efficient way to discover drugs. It may be an ethically unpalatable fact and something you would never wish to re-visit, but I think probably bits of it could be revisited with not huge risk."

Some of the most successful drugs are the result of serendipity, as is the case for metformin, the most widely used drug for type 2 diabetes. Decades ago a researcher in the Philippines studying an obscure compound to treat flu and malaria reported that it also seemed to lower blood sugar. In 1957, a Parisian scientist noticed that published observation and tried the drug in animals. British researchers tried it in diabetics in the 1960s, and it worked surprisingly well. The original compound was discovered as an herbal extract, and to this day scientists don't understand the biological mechanism. But it doesn't matter. It works.

Chance still plays a major role in drug development. Doctors monitoring their patients do occasionally make surprising discoveries. Patients taking minoxidil for blood pressure control noticed unusual hair growth—and Rogaine was born. And with more than 1,000 approved active ingredients in our drugs, surprises (sometimes pleasant) are inevitable. "Most new uses of drugs are developed by doctors through field observation," Scannell told me.

The lesson here is twofold. First, it's better not to assume that a specific drug works with pinpoint precision. Most drugs "are magic shotguns, not magic bullets," he said, and sometimes the "off-target" effects can be useful (clearly, they can also be nasty side effects). Second, it's important to remember that many important discoveries start with human beings in a medical clinic rather than with mice in cages.

The bottom-up approach to drug development isn't all doom and gloom. There's a growing list of successes around a new generation of drugs known as biologicals. These are antibodies or other molecules designed to hit very specific targets, and they have on occasion succeeded in translating basic biological insights into worthwhile treatments. Checkpoint inhibitors are one example, raising hopes that targeting pathways in the immune system to fight tumors might finally control cancer. The progress isn't turning out to be as dramatic as proponents had hoped, and even the drugs that work do so for a minority of patients. But these are early days.

Animals will still be an important element of that research, despite their shortcomings. Model organisms still provide valuable insights into fundamental biology—whether it's tetrahymena for its revelations about telomeres, fruit flies for their insights into genetics, or even mice used to study the basic wiring of the brain. If researchers studied only human beings, those insights would be vastly more difficult to achieve, not to mention ethically impossible in many instances.

Even so, translating that basic biology into medical advances is anything but straightforward. And that is even more the case for another ubiquitous tool used in biomedical research: disembodied cells floating in laboratory flasks. A rich history shows that this type of research is rife with problems. But dire warnings have largely been ignored. That is one of the starkest cautionary tales involving rigor and reproducibility in biomedical research.

Chapter Five

TRUSTING THE UNTRUSTWORTHY

IN 1994, NINA DESAI and her colleagues at the Ohio State University published some good news: they announced the creation of a new tool to produce test-tube babies. The team said it had isolated a line of human cells from a woman's womb and coaxed them to grow perpetually in the lab. They planned use these cells as microscopic nursemaids, to help provide growth factors for human embryos being nurtured in the lab to help infertile couples conceive.

Desai moved to Cleveland and in 2000 became director of in vitro fertilization (IVF) research at the Cleveland Clinic Foundation. There she put her special cell line to work. As she described in a 2008 paper, she would put 15,000 to 30,000 of these cells in a plastic dish. A permeable membrane kept them from direct contact with the embryos but allowed the biological materials they produced to wash into a second chamber, which contained embryos being nurtured in preparation for implantation in a woman's womb. She reported that the procedure had been used to treat the embryos of 316 women between January 2004 and March 2007. By the date of her publication, the tests had resulted in 111 apparently healthy infants born to 76 women.

But the story was not quite so simple. When Christopher Korch, at the University of Colorado School of Medicine in Denver, came across this paper, he was puzzled about the exact identity of the cell line that Desai said she had established. Normal cells can be coaxed to grow in the lab for a while, but they eventually peter out. This cell line kept going—it had become immortal. "When you hear that, then you really have to worry," Korch said. "That is the warning sign you've got something else in there." Cells rarely transform spontaneously, but they readily become contaminated with aggressive cell lines that can easily move through a lab. This enormous contamination problem dates back more than half a century and has cast doubt on a large slice of the biomedical research literature.

Scientists first managed to keep human cells proliferating in the lab in 1951. Researchers at the Johns Hopkins Hospital extracted some cervical cancer tissue from a woman whose story is chronicled in Rebecca Skloot's *The Immortal Life of Henrietta Lacks*. Those rapidly proliferating cells became a favorite lab tool for scientists interested in studying cervical cancer in particular and human biology more generally. Unlike typical cells extracted from a person, these kept dividing and proliferating indefinitely. This immortal cell line, labeled HeLa, was just the first of many. And because HeLa cells grew so quickly, they became rapacious weeds in the world of biomedical research labs. The slightest lapse in hygiene can transfer a HeLa cell from one dish to another that's harboring a different line. The fast-growing HeLa cells quickly crowd out the other cells and simply take over.

Through the 1960s and 1970s, Walter Nelson-Rees made no end of enemies in science by testing cell lines purported to be from many different cancers and pointing out correctly—but brusquely—that they were in fact HeLa cells. Nelson-Rees, who curated the National Cancer Institute's cell repository in Oakland, California, waged a bitter campaign at that time to make scientists realize that they were not, in fact, exploring breast cancer or liver cancer or whatever they thought they were working on. Word of his campaign spread throughout the science world and far beyond into the newspapers and magazines of the day. In 1986, science writer Michael Gold wrote *A Conspiracy of Cells*, a lively history of Nelson-Rees and his campaign against HeLa. And how did the scientists using cells in their research respond? They mostly ignored the problem.

Even today, despite the easy availability of conclusive identity tests, HeLa crops up frequently in labs intending to investigate something else. KB cells, used extensively to study oral cancer? They're actually HeLa. Human epithelial type 2 (HEp-2) cells, thought to be a sample of cancer from the larynx? Actually HeLa, as well. Chang liver, Intestine-407 (Int-407), and WISH cells? HeLa, each and every one. That's ancient history—all were exposed for what they are in the 1960s. Even so, more than 7,000 published studies have used HEp-2 or Int-407 cells, unaware that they were actually HeLa, at an estimated cost of more than $700 million.

And that's just a sliver of the problem. A 2007 study estimated that between 18 and 36 percent of all cell experiments use misidentified cell lines. That adds up to tens of thousands of studies, costing billions of dollars. About a quarter of those misidentified lines are actually HeLa, but there are plenty of other masqueraders out there. Sometimes even the species isn't correct. Nelson-Rees found a "mongoose" cell line was actually human and determined that two "hamster" cell lines were from marmosets and humans, respectively. "Have the Marx Brothers taken over the cell culture labs?" Roland Nardone asked in a 2008 paper bemoaning this state of affairs.

Nardone, a biologist at the Catholic University of America, took up the cause of contaminated cells in 2005, after his son wished him a happy seventy-seventh birthday and asked what he intended to do with the rest of his life. After some reflection, he decided to pick up where Nelson-Rees had left off. In 2015, at the age of eighty-seven, Nardone, with a shock of white hair and bushy white eyebrows, rose out of a wheelchair and gripped a lectern at the National Institutes of Health (NIH) to talk about his ten-year quest to straighten out this glaring problem. He had written a paper in 2007 urging zero tolerance for these bogus cell lines. "I thought the bandwagon would be crowded with all the people who would want to jump on board," he said. That was not to be the case. The NIH actually responded to Nardone's letter promptly and set out a new policy. "But it wasn't strong," he said. "It just encouraged increased vigilance and oversight." *Nature* published an editorial in 2009 declaring that it would institute a new policy to set this right but in fact took no formal action for years. By the end of his talk, Nardone had worked himself into a state of righteous indignation. "How dare we not authenticate our cells, regardless of what we're doing!"

Nardone was not the only scientist growing alarmed about this issue. Gradually a group started to coalesce. Amanda Capes-Davis, an Australian cell biologist, emerged as a leader of this loose-knit band. She had set up a cell bank at her institute in Sydney and quickly became aware that many cells circulating among scientists were misidentified. She cast around to see if anyone had compiled a list of those imposter cells and couldn't find one. So in 2009 she quit her job to devote her full attention to this matter. "I spent six months at home, working on what I thought was a reasonable list," she told me. When she got about two-thirds of the way through, she discovered that a Scottish cell bank scientist, R. Ian Freshney, had been working on a list as well. "My first reaction was that six months was wasted. Someone else has already done this," she said.

"My second reaction was, let's look at the lists." It turned out that although they shared a lot, each scientist had dug up unique problem cell lines from the scientific literature. A partnership was born.

Their list of contaminated cell lines gradually grew into the hundreds. People had various methods of identifying these wayward cells. Nelson-Rees in the 1970s and 1980s examined them under a microscope to look at the patterns of the bands on their chromosomes. He also ran some enzyme tests that were more informative but still not definitive. By the 2000s, however, scientists could readily use genetic fingerprinting techniques to identify cell lines. So the nation's research cell bank, the American Type Culture Collection (ATCC), decided it was time to settle on a standard test that biologists around the world would use to verify their cell lines. Officials there asked Capes-Davis, Freshney, and a small band of their associates to sit down and write that standard. It took two years, but in the end they settled on a cell-fingerprinting technology that was reliable, reproducible, and inexpensive—typically less than $200 per test.

For Capes-Davis, this whole issue became a labor of love that she performed as a volunteer from the studio of her Sydney home. Her passion was partly intellectual—there are intriguing mysteries to solve here—and partly ethical. "When I tried to establish cell lines [as a researcher] I went to collect samples from potential donors," she said. "We have a responsibility to look after those samples." With the standard in hand, Capes-Davis was then anointed to chair an organization that sprang up around this issue, the International Cell Line Authentication Committee. It maintains and updates the list of corrupted cell lines, which by 2016 had grown to 438, with no end in sight. And she continues to expose the history of these wayward cell lines—particularly the ones that Walter Nelson-Rees was already flagging decades ago, such as KB and Chang liver. "Those are my horror stories because they are still so widely used," she said. "These are from the 1960s—why are we still using them so much?"

One of the most flagrant examples that Amanda Capes-Davis, Christopher Korch, and their colleagues investigated involved a cell line widely used to study breast cancer. This story starts in Houston, on January 23, 1976. A thirty-one-year-old woman diagnosed with early-onset breast cancer was seen at the MD Anderson Hospital and Tumor Institute. Fluid had been accumulating around her lungs. A hospital worker drew some into a syringe and delivered it to the laboratory of Relda Cailleau. Cailleau and her colleagues were in the midst of a six-year project to capture breast cancer cells in order to cultivate them in the laboratory. The cells from this young woman did indeed take hold in a petri dish, becoming part of a collection of nineteen different breast cancer cells extracted between 1973 and 1978 at MD Anderson. The cells from this particular woman were dubbed MDA-MB-435 (and sometimes labeled MDA-435).

And it turns out they were especially useful, as they had the rare ability to spread in mice the way cancer metastasizes in people. In short order, labs around the country clamored for samples of MDA-MB-435 to study metastatic breast cancer. It proved so popular that in the late 1980s, the National Cancer Institute selected it as one of sixty key lines that would get extraordinary attention. This collection, dubbed the NCI-60, would be used to test hundreds of thousands of potential new cancer drugs. Over the years, hundreds upon hundreds of journal publications reported breast cancer experiments involving MDA-MB-435, as scientists hoped they were homing in on better treatments or even a cure. It turned out that MDA-MB-435 was an imposter.

The cell was unmasked quite by accident. Back in the late 1990s, scientists at Stanford University were developing a test that would allow them to look at a biological sample and see which genes are switched on or off in any given cell. Doug Ross was a postdoctoral researcher in a star-studded laboratory that helped develop these powerful new genetic tools. His boss, Pat Brown, put him in charge of a marquee project: a study of all sixty of the lines in the NCI-60. He and his colleagues set up an experiment to investigate about 8,000 genes in these cancer cells and to look for patterns. Which genes were turned on? Which were turned off? How did they differ from one type of cancer to the next?

In March 2000, Ross and his colleagues reported exciting results. Using their powerful new technique, they could tell one type of cancer from another simply by looking at patterns to see which genes were active and which were silent. The various lung cancer cells included in the NCI-60 had one genetic pattern in common. Prostate cancer cells all shared another. Melanoma cancers had their own unique gene-expression fingerprint. And so did breast cancer cells—well, almost all of the breast cancer cells. MDA-MB-435 didn't come out looking like a breast cancer. Its gene pattern matched the melanoma cells and "really had nothing to do with the breast cancer cell lines," Ross told me. "So we repeated the experiment to make sure we didn't screw it up"—and got the same melanoma pattern. Ross borrowed a different sample of MDA-MB-435 from colleagues at Stanford. Same thing. It was looking a lot like a melanoma. "We just mentioned in the paper the possibility its tissue of origin was misidentified," he said.

Further investigation has since revealed that the cells are nearly identical to another cell line in the NCI-60, a melanoma cell line called M-14. The NCI put up a note of caution to alert breast cancer researchers that the cell line appeared to be misidentified. Some scientists who had spent many years studying this "breast cancer" dug in their heels. "People were very invested in the tremendous effort they'd put into the cell line," Ross said. Some developed a convoluted rationale to explain how MDA-MB-435 could still conceivably be breast cancer cells—an argument that holds little sway in the field. "You just shrug your shoulders and say, 'That seems very unlikely to me,' but that's what people want to believe," Ross told me. Many scientists still don't realize that this is a melanoma cell line, and they continue to publish "breast cancer" studies based on this skin cancer cell. There are now more than 1,000 papers in scientific journals featuring MDA-MB-435—most of them published since Ross's 2000 report. It's impossible to know how much this sloppy use of the wrong cells has set back research into breast cancer.

Christopher Korch was fascinated by this story and spent many weeks, along with Capes-Davis, figuring out exactly what happened. Korch, now retired from academia, spends his energy, like Capes-Davis and Roland Nardone, trying to untangle decades of bungled science surrounding cell cultures. Among other investigative work, he has been trying to figure out whether there were any original, unadulterated breast cancer cells correctly labeled as MDA-MB-435. In the course of that detective work, Korch found a 1979 student dissertation that alludes to a collaboration between Relda Cailleau in Houston and Donald Morton in Los Angeles. Morton had isolated M-14—the melanoma cell line—a year before the thirty-one-year-old patient donated fluid containing her breast cancer cells. Korch suspects that the Houston cell lines got contaminated when Cailleau visited Morton's lab.

Korch told me that he spends hours every day poring over these old histories—not just for MDA-MB-435 but for many other lines. It's partly an intriguing detective story, but Korch also wanted to measure the magnitude of the problem. He started with the list of known contaminated cell lines. For example, Intestine-407, which is actually HeLa, has been used in at least 1,300 published experiments. HEp-2, also HeLa, is used in 5,700 papers. All told, he figured perhaps 12,000 papers are based on bogus cell lines. But that's not the end of it. He estimates that, on average, each of those papers was cited in other papers thirty times. "When you start doing the multiplication, we're talking about billions of dollars that have been spent using a cell line inappropriately."

Now, that's not completely wasted effort. Michael Gottesman at the National Cancer Institute had acquired KB cells in the 1980s from the national cell bank (the American Type Culture Collection) and used them for some of his studies to figure out why tumors develop resistance to anticancer drugs. When KB was unmasked as HeLa, Gottesman wasn't exactly thrilled, but he says in his case it actually didn't matter. "We were not particularly interested in the origin of the tumor," he told me. "We just wanted a cancer cell line. It had the properties we wanted." They grew fast and were relatively sensitive to anticancer drugs. And Gottesman was able to extract a gene from them to move his studies forward. "Even though there have been problems [with misidentified cell lines] they don't always torpedo the research," he said.

Korch agreed that research based on contaminated cells isn't a total loss. "But how do you sort the wheat from the chaff to find what is usable?" It's no simple matter even to identify tainted studies. For example, a search for "KB" in medical databases will inundate you with papers that use "KB" as the abbreviation for "kilobase," a word used all the time in genomics studies. "I don't see that the literature is going to be cleared up, ever." Korch said. "It's a gargantuan task." After talking for more than two hours about these issues, I asked him if dealing with these problems is a passion or an obsession for him. He paused and stroked his white beard. "Where's the line?" he replied with a smile. "I suppose I'm more of an obsessive person."

That obsession ultimately led him to the papers describing Nina Desai's work at the Cleveland Clinic. Korch couldn't at first figure out what the cell was, because Desai never called it by name in her published papers. But then he came across an abstract from a scientific meeting in which Desai and colleagues from Emory University referred to a human endometrial cell line by name: EM-42.

Korch eventually tracked down a sample of EM-42 cells and confirmed his worst fears. They weren't healthy human cells after all—they were HeLa. He couldn't tell for sure if EM-42 was exactly the same cell line as the one that Desai used in her fertility clinic. Perhaps Desai used another cell line or contamination with HeLa occurred later. Scientists usually resolve these questions by sharing ingredients and talking to one another. But Korch said Desai had not replied to his e-mails and phone calls to resolve the story.

If EM-42 was used and the fertility work had been conducted flawlessly, the membrane would keep the HeLa cells away from the developing embryos. But labs do make mistakes. And the membrane wouldn't stop snippets of DNA, or indeed viruses, from moving from the cells to the human embryo. Did the scientists at the fertility clinic know they may have been using cancer cells? And what about the parents? Korch asks. Of all the examples of potentially misidentified cell lines, he said, "that is the scariest I've seen."

The use of cancer cells in IVF is also far outside the comfort zone of Andrew La Barbera, speaking as chief scientific officer of the American Society for Reproductive Medicine. He had worked in an IVF lab earlier in his career and would never have considered growing human embryos on a cancer cell line. "We would have considered that to be fraught with hazard." Regardless of their precise identity, the simple fact that these cells were proliferating endlessly suggested something was wrong with them. "I don't know how you would prove that any cell line would be harmless."

Fertility clinics have waded into uncertain waters before. Some doctors tried to help human embryos grow by mixing in cells from monkeys or cows. In 2002, the Food and Drug Administration officially frowned on those procedures, given the potential hazard of a virus infecting the embryo. Desai's 2008 paper reported that the babies were all healthy. As of this writing, the children would be about ten years old. La Barbera said their mothers should be informed of a potential laboratory mix-up, just in case. But if the Cleveland Clinic and Desai believe that as well, they aren't saying. Desai would not discuss this story with me, and the Cleveland Clinic would not comment or talk about its processes for protecting the patients in these experiments.

Contaminated cell lines skew many different lines of research—including studies of the brain cancer glioblastoma. More than seven hundred studies report using a cell line called U-373, originally isolated as a glioma cell. For example, a study in Belgium used U-373 cells to argue that an experimental drug called ISO-1 could be worth testing as a treatment for brain cancer. Unfortunately, many of those studies may have been wasted effort. Cell banks like the American Type Culture Collection—which are supposed to be authoritative sources of reliable cell lines—made U-373 widely available for research. But in 1999, scientists discovered that people ordering U-373 were actually getting another cell line, U-251. At first blush this didn't seem like that big a deal. U-251 also happened to be a glioma cell, so at least scientists using U-373 were still studying the right disease. But in 2014, scientists in Norway took a closer look and were very unhappy about what they saw. It turns out the U-373 cells were not merely mislabeled; they were a strain of U-251 that had been circulating for many years and had accumulated so many mutations and other genetic alterations that they barely resembled glioblastoma at all. There was no telling whether these cells had any relevance at all to human cancer.

Cell banks eventually tracked down an early sample of correctly labeled U-373 cells, and after 2010 they once again made those cells available. But many researchers don't bother to buy fresh cell supplies from a cell bank when they're starting a new round of experiments; they may pull out an old supply from a freezer or borrow them from a colleague down the hall. The result? Scientists continue to publish studies all the time involving "U-373" cells, "in which it is not obvious if the authors have used the cross-contaminated U-251 or the correct one," Anja Torsvik at the University of Bergen in Norway and her colleagues noted in a 2014 article.

More than 1,700 papers have been published using what is arguably the most commonly used glioblastoma cell line, U87. And it turns out that it is troubled as well. Biologists in Uppsala, Sweden, isolated it from a forty-four-year-old woman with brain cancer nearly fifty years ago and managed to grow it as a perpetual cell line in their lab. In 2016, scientists from Sweden decided to compare the original cell line from their freezer with the U87 cells that have been sold by ATCC and used widely around the world. It wasn't a match. Somewhere along the line, an imposter took over. In fact, the widely used U87 has a Y chromosome, so it appears to have come from a man. Fortunately the imposter is also a brain cancer. But the episode shows that even cell lines that have been validated by cell banks can still be misidentified.

It's not clear how much value comes from research that relies on cell lines in the first place. Much as scientists appreciate the convenience of studying a disease in a petri dish, the results are often hard to apply to human illness. The very act of propagating cells in the laboratory changes them profoundly. In the first place, the process typically selects for cells that thrive while clinging to a plastic dish in a single layer. The cells are exposed to normal atmospheric oxygen levels, which are about four times higher than cells encounter in a tumor. "A lot of the regulatory factors that affect the growth of tumors is oxygen regulated so it's a huge difference," said Michael Gottesman at the NIH. These cells grow far more rapidly than they would in a tumor. In fact, cell lines derived from all sorts of cancers end up looking much more like one another than they do the original tumors from which they came. So the differences between the cell in a tumor and its progeny in a plastic dish can be quite dramatic. "Some people say that HeLa is a new species," Gottesman told me. "It has lots of human components, but the cell line is so evolved. The chromosomes are all rearranged. It survives in tissue culture. It grows well. So it has made all these changes to adapt" to the environment where it now makes its home.

Gottesman said there is still useful information to be gleaned from cancer cells growing in a plastic dish, but it turns out these experiments typically don't have much direct relevance to treating human tumors. The NIH established the NCI-60 in the late 1980s with high hopes that drugs showing promise when tested against these cells would also work for tumors in people. "I think it's been a great disappointment," Gottesman said. "It basically didn't pan out." He and some colleagues looked back on the many years of experiments, searching for a drug that resulted from that massive effort. They found just one: Velcade, a treatment for multiple myeloma (a cancer of immune system cells). A few months after we spoke, the cancer institute terminated the NCI-60 program. It's launching a new technology it hopes will be better.

It's easy to avoid the ubiquitous problem of misidentified cell lines. Scientists should simply ship a sample of their cells off to a commercial testing lab before they start their experiments to make sure the cells are what they expect. They should also authenticate their cells the same way after the experiment is done. Scientific funding agencies and journal editors are gradually pressuring scientists to do just that, but as Nardone discovered a decade ago, authorities are reluctant to insist. For one thing, scientists are independent operators and don't like being told what to do. For another thing, the tests are not free, and even a couple hundred dollars can seem like a lot to a lab struggling to make ends meet. That penny-wise-but-pound-foolish attitude is unfortunately part of the culture of academic science, and as long as the consequences for a scientist's career are minor, there's not a great deal of incentive to change.

Many for-profit researchers see things quite differently. They simply can't afford to be wrong about their cell lines.

With that in mind, shortly after scientist Richard Neve arrived at Genentech, a biotech giant on the muddy shore of San Francisco Bay, he soon found himself immersed in a project to make sure the company was using only clean cell lines. He showed me around the sun-drenched campus in early 2015 (before he decamped to a different company) and led me to three gleaming tanks filled with supercold liquid nitrogen. These constitute the Genentech cell bank. Inside are nearly 100,000 plastic vials, each containing a tenth of a teaspoon of frozen cells. Neve said about 1,800 separate cell lines are stored here. Each morning, technicians fill requests from the company's scientists by fishing out individual vials, scanning their barcodes to make sure that they have the right one, and then sending them along to the research scientists (Neve included) at the company. Anybody at Genentech who wants to use cells has to start here—borrowing from the neighbor down the hall, common in academia, is verboten. This assures that the cell line is correct. The company cell bankers also test regularly for a bacterial contaminant, mycoplasma, which is a major headache. It shows up regularly in academic labs and completely throws off the results of experiments.

"We're trying to avoid any sharing of cell lines because you just don't know what you're getting," explained Neve, a Brit with a supershort crew cut and the exacting manner of a neat freak. Genentech routinely sends cells out to be authenticated with the standard commercial test. Neve and his team also developed an in-house testing system using a different technology, called SNP analysis, which costs just $6 per sample in ingredients (of course they had to invest in a laboratory instrument as well). Neve takes this issue very seriously. In fact, he is part of the loose-knit Capes-Davis group. He has published data in scientific journals to help scientists identify bad cell lines. The company's cell-handling operation is highly automated—with robots to move materials around and scanners to keep track of everything. Unlike academics, who are rewarded for publishing intriguing results, companies only benefit if the research ends up delivering a profitable product. Errors cost time, and time is money. And their systems do catch potentially serious problems. "If we [Genentech] are making mistakes," Neve told me, "God knows what it's like out there" in the world of academia.

Tests to authenticate cell lines aren't a panacea. For example, they can't identify whether a given cell is from a sample of liver, brain, or gut. That's an increasing concern because scientists are isolating new cell types all the time and using them in place of the more traditional immortalized cancer cells. In addition, there are no routine tests for identifying cells that come from laboratory animals, and those too are being used with increasing frequency. So even if scientists adopt the current authentication technologies, they will need to be developing new tests to keep up with the trends in biomedical research.

And as if cell line problems aren't bad enough, there's an even bigger problem involving another ubiquitous laboratory tool: antibodies. Monoclonal antibodies are custom-built molecules designed to identify and glom onto specific materials inside cells. When they work, they're great: just as your natural antibodies can target a single molecule on the surface of a specific germ, laboratory-produced antibodies are supposed to work like guided missiles, to home in on a specific substance. They often carry a fluorescent marker so biologists can easily flag the material they're looking for. Antibodies are quite reliable in some circumstances—for example, in early pregnancy tests, where they tag a hormone that's produced during pregnancy. But far too often, the 500,000 antibodies marketed for research by a multi-billion-dollar industry don't work as advertised. Many are produced by injecting the targeted substance into a rabbit and collecting the antibodies that result—a technique highly prone to providing misleading results. Biomedical researchers have been slow to grasp just how big a problem this is.

Stan Artman's story shows what's at stake. In the fall of 2011, the fifty-two-year-old Atlanta resident looked down at his leg and saw a black spot. "I had been golfing that day, and I was in the woods chasing balls. I thought maybe I'd picked up a tick." He tried a couple of tick-removal tricks but the black spot remained. His wife, as it happens, is a dermatology nurse, and she sent him off to a doctor for a closer look. As Artman tells the story, the dermatologist at Emory

University wasn't sure what the spot was. Artman decided to get it removed just in case. Pathologists still couldn't figure out whether it was worrisome, so they sent a sample off to the University of California, Los Angeles, for further analysis. The word came back: it was probably melanoma, a potentially lethal form of skin cancer.

Melanoma poses a real conundrum for patients, particularly less advanced cases like Artman's. Surgery can be curative, but patients can also improve their odds a bit if they sign up for a year of unpleasant and sometimes debilitating interferon treatment. For Artman's stage of disease, it only benefits about one in every thirty patients who take the plunge, and there's no way of knowing in advance who that one will be. Instead, Artman's doctors recommended that he try an experimental antimelanoma vaccine and have more extensive surgery to remove lymph nodes where the cancer could conceivably be hiding. Watchful waiting was an option, but he said doctors told him, "If you're the type of person who a year from now developed a mass in your lung, you might feel you've missed your golden moment to get everything out."

He opted for surgery. A few days later, he noticed with alarm that the surgery site had turned bright red. It was "a heartbreaking moment, of 'now what,'" he said. "What the hell is this?" It turned out to be an infection that sent him back to the hospital for nine days of intensive antibiotic treatment. So even the seemingly conservative step of surgery carried risks, while the rewards were very uncertain.

"There's this big gray area when it comes to melanoma," Artman said. A blood test to guide some of these difficult decisions could have given him some idea of whether he would remain cancer free. "It would take some of the question out of it," he said. "It wouldn't take the worry out if I came back positive, but I'd feel better knowing something was definitely one way or the other."

David Rimm at Yale University was apparently well on the way to developing a test like that, and he might have succeeded had he not had a very unfortunate encounter with bad antibodies. A renowned pathologist, Rimm realized he might be able to develop a test using commercially available antibodies to identify melanoma patients who would benefit from additional, aggressive treatment. So he collected tissue samples from about two hundred patients, some with metastatic melanoma and some with less aggressive disease. He then tested out about eighty different antibodies purchased from various companies to see if an antibody combination would identify patients more likely to benefit from interferon, more surgery, or other potentially risky treatments. The antibodies were all directed at known features of melanoma cancer cells. No single antibody provided a strong signal, but Rimm found that if five specific antibodies all found their targets and lit up, that pattern appeared to be a strong predictor of patients who would most benefit from aggressive treatment. "So we had ourselves a test. We were psyched."

Rimm wanted to make sure he got the same results with a second group of patients, so he applied for two grants from the NIH to continue the work. Reviewers were ecstatic with the proposal, giving it a top score. Rimm got two grants for $1 million each to pursue the work. He ordered fresh antibodies from the suppliers to start his confirmatory experiments. That's when the project started to unravel. When he ran the same tests on a different sample, three of the five antibodies still lit up as expected. But the other two did not. And the three alone weren't reliable enough on their own as a test for cancer patients. What went wrong? "To this day we don't know what it was," Rimm said. His best guess was that the new batch of antibodies he ordered weren't exactly the same as the initial batch. The reason could be as simple as that the first batch of antibodies had come from one particular rabbit and the second from another. Whatever the cause, it was a fatal error for his would-be test.

After many years of increasingly excited effort and millions of dollars in research funding, the whole thing fell apart. It still seemed like a great idea, but "we would've had to start from scratch to reinvent our test," Rimm said. He was demoralized and defeated. "I didn't see how we could possibly fund this work again." He could just imagine what the review committee would say if he submitted another funding request: You tried. You failed. It's over. So he stopped working on melanoma altogether.

Rimm said he had not realized how unreliable antibodies could be. He had assumed that they were as trustworthy as anything else he bought from a lab supplier and that he could simply believe what he read on the label. But, especially when produced in live animals, antibodies are anything but dependable. While in theory they bind to a unique site, they may also glom onto several different proteins, not just the one that scientists expect. And it's not always a simple matter to identify these "off-target" effects. As a result of this painful experience, Rimm has become an evangelist for cleaning up the mess with antibodies. And it's a big mess. Glenn Begley said faulty antibodies were apparently responsible for a lot of the results he was unable to reproduce at Amgen.

Rimm's experience is just one example of the trouble with antibody tests. Another mix-up cast doubt on the very existence of a hormone that helps burn fat when people exercise. "Irisin," discovered in 2012 by Bruce Spiegelman and his colleagues at Harvard, appeared to turn standard body fat into "brown fat," which actively burns calories and may play a role in weight loss. Of course, anything to do with a potential fat-burning pill is immediately interesting. Scientific supply companies sprang into action and started selling antibodies that they said were specific for identifying irisin. Soon dozens of scientific researchers were using those antibody tests to see whether exercise, diet, or even Turkish baths would affect irisin levels in people.

Harold Erickson, a biochemist at Duke University, became skeptical about irisin. Though it was not his field of study, he was drawn to what he saw as inconsistencies in the story. He teamed up with a scientist in Germany who had tried—and failed—to find irisin in horses that had just finished heavy bouts of exercise, which is exactly when biologists would expect to find the hormone circulating in the blood. Erickson ordered some of the antibody kits that target irisin and concluded that they might not do so at all. He published a paper that went so far as to call irisin a "myth." (Irisin is named for a Greek goddess, so wordplay was irresistible.) "That was pretty aggressive" as a choice of words, Erickson admitted to me. He also encouraged his university news office to put out a fairly aggressive press release, touting his assertion that irisin may in fact not exist.

That infuriated Bruce Spiegelman at Harvard. After the public challenge, he and his colleagues went back to run a new set of experiments that were much more sensitive than the off-the-shelf antibody tests. They published a follow-up paper announcing that they did indeed see irisin circulating in human blood. The levels were just very low. That's not unusual for hormones, which can be active in small concentrations. But those low levels can skew antibody tests.

Erickson argued that irisin's apparent blood levels are so low that commercial antibody kits shouldn't be able to detect it. (Indeed, at least one company withdrew its product from the market during this flap.) I asked Spiegelman whether he thought those kits were valid. "I have no idea," he said. "That's somebody else's problem. We've never used them."

Though he still harbored doubts, Erickson backed away from his claim that irisin doesn't exist. But the entire episode has left the field in a bit of turmoil. Only a few researchers in the United States have received federal grants to fund work on it, even though, if it pans out, it could have big consequences for obesity, diabetes, and other major diseases. Researchers overseas are still publishing on it and still using the disputed antibody test kits. Drug companies could also be working quietly on irisin, with an eye toward a weight-loss pill. "They wouldn't necessarily tell me," Spiegelman says. Even though they would ultimately have to license Spiegelman's patent for irisin, he says they wouldn't bother to ask until they were ready to try the drug in people.

Spiegelman was quite confident of his results, but it will take some time for the dust to settle, as scientists argue the facts of this provocative finding. He said that's simply a consequence of working on the cutting edge of science. "If you don't want to deal with these things, work on something that's been done 100 times."

Often scientists either don't realize that they've run into problems with their antibodies or simply heave a sigh and move on to a different project. But David Rimm decided to call attention to bad antibodies by publishing a paper flagging this as a major problem. The paper has been downloaded more than 35,000 times, "so there's hope that people actually want to do something about this," Rimm told me. He's been working on a solution. He led me upstairs to one of the labs he oversees. Here, scientists were at work constructing microscope slides, called index arrays, dotted with tissue samples smaller than the period at the end of this sentence. One slide had nearly one hundred dots. The idea is that each slide can be a miniature laboratory to validate antibodies. Put a drop of antibody on each slide, and some of the dots should light up; others should not. Dots with a small amount of the target material should light up a little; dots with a lot of it should light up brightly. Rimm doesn't trust his own eyes to tell the difference (that could lead to observer bias), but there are instruments that can measure all this precisely. "When you use people [to eyeball a sample], you're just destined to fail," Rimm said. "I like people. I am one. I like pathologists. I am one. But it's the wrong tool for the job."

He helped develop and commercialize some of those precision instruments that are now sold to research labs. Rimm is also trying to get antibody companies to construct and sell index array test slides as well. That would go a long way toward solving the antibody problem, but at a high cost: his lab spent more than two months and $10,000 to produce a slide that would check the performance of a single antibody. To make a slide like that for every one of the 500,000 antibodies out there would run into the billions of dollars.

In 2014, Rimm raised all the problems with antibodies at a scientific meeting at the St. Johns Laboratory in London. After he spoke, John Mountzouris, from a big antibody company called Abgent, said that his company had recently run a very basic test of all 18,000 antibodies in its catalog. "As a result several thousand antibodies were discontinued immediately," Mountzouris told the gathering. It was a moment of reckoning for his firm. "The revenue did decline from antibody sales. We hope that's short term," he said. But he assured higher-ups in the company that their scientist-customers will eventually realize that they're better off buying validated antibodies than going to any of the dozens of other companies that sell products with similar labels but less rigorous verification (often at a lower price).

Not only did the company whittle down its existing catalog; it dramatically slowed the rate of introduction for new antibodies, from 4,000 or 5,000 to 1,000 per year, Mountzouris told the meeting attendees. "That will allow us to produce antibodies that we will be proud to put out there for people to buy."

The British company Abcam is by far the largest in the business (though because there are so many players, it accounts for only 20 percent or so of the market). It, too, has been amping up its systems to weed out untrustworthy antibodies and to sell based on quality rather than price. Abcam officials told me the company has taken basic steps to validate almost all of the antibodies it sells. In 2015 Abcam also started screening antibodies by removing the purported target of the antibody from cells, using a gene-editing technology called CRISPR. If those antibodies still light up when exposed to cells that don't contain their intended target, they are obviously not working as expected. The company plans to validate five hundred antibodies a year this way.

The first several hundred tests showed that 60 to 70 percent of the antibodies passed this stringent test, Alejandra Solache, Abcam's head of reagents, product development, and manufacturing, told me. That means, of course, that the remaining 30 to 40 percent of these products—many of them popular antibodies used widely in research—weren't up to snuff. "When something doesn't pass we will remove it from the catalog and will notify the customers that basically this particular antibody is not specific for this protein," Solache said. The company sells well over 100,000 products, mostly antibodies, so this expensive and time-consuming validation effort will only apply to a tiny percentage of its products for years to come.

Antibody companies like Abcam that are willing to invest in improving the quality of their products can help put a dent in the reproducibility problems brought on by poor antibodies, but Bill Campbell, Abcam's general manager for the Americas, said commercial enterprises can't solve the whole problem. When he worked in a lab, he said, he always ran control experiments to assure himself that he wasn't falling victim to imperfect materials. "Scientists also need to make sure they do the right controls and not try and advance things too quickly by taking shortcuts," he said.

It would help if scientists working with antibodies could refer to standard laboratory procedures to help them avoid some of the common pitfalls. David Rimm is pushing for that now. Unfortunately, standards for antibodies aren't nearly as straightforward as standards for authenticating cell lines. Experiments involving antibodies are more varied, so one size does not fit all. And as the story of cell lines makes clear, simply having a standard isn't enough. Most scientists must be coerced by funding agencies or their employers to run these tests.

Assuming scientists do step up to take on the problems of authenticating cell lines and antibodies, that could address perhaps a quarter of the problems underlying rigor and reproducibility issues. That's a big chunk, to be sure. Improving experimental design would further reduce these unforced errors in science. But carefully designed and executed experiments are still worthless unless their results are analyzed with care. That's the next crucial link in the chain of scientific rigor.

Chapter Six

JUMPING TO CONCLUSIONS

RARELY IS A scientific discovery so galvanizing that Congress passes a resolution calling for more funding of the research. But Congress did just that in 2002, on the heels of an exciting announcement. Researchers at the Food and Drug Administration (FDA) and the National Institutes of Health (NIH) trumpeted a new test they'd developed to detect ovarian cancer, which up until that point could only be diagnosed with surgery. Many women learn they have the disease in its later stages, when it's especially hard to treat, while others undergo unnecessary surgery simply to rule it out.

News of this putative new test, based on a novel technology, made headlines. The researchers who discovered it ended up on the *TODAY* show. Scientists were excited as well. Keith Baggerly and his colleagues at MD Anderson Cancer Center in Houston started scrambling to assemble a lab that would let them do this kind of testing. If it worked for ovarian cancer, they reasoned, it seemed likely that the technology could diagnose many other forms of cancer in early, more treatable stages. And the excitement arose in part because this was no ordinary blood test. Here's the novel concept: scientists extracted proteins from blood samples and put them into a device called a mass spectrometer—basically a sorting machine that separates molecules by mass. The original researchers compared results from fifty women with ovarian cancer and fifty healthy women. They reported seeing a notable difference, identifying a pattern in the cancer cases most of the time. This idea marked the first step into an exciting frontier. Instead of searching for one particular molecule as most lab tests do, this test looked for a broad pattern, a protein spectrum. It seemed like the start of something big. Really big.

Baggerly naturally wanted to see if he could also see the pattern in the data. "We looked at it fairly extensively for a few months," Baggerly said, "and we couldn't find the patterns that they were reporting." Other scientists started raising doubts as well, commenting on the paper's methods and conclusions in short letters to the *Lancet*, which had published the original research. Baggerly kept gnawing on this bone and eventually realized that the data he had been analyzing from the original report had already been cleaned up a bit. When he went all the way back to the raw data, lo and behold, he did see a significant difference between women with and without ovarian cancer.

But there was a problem. The difference Baggerly saw was in data that scientists generally throw away because it's untrustworthy. It reflected "noise" generated by the machine, he concluded, and had nothing to do with detecting a real protein fingerprint for ovarian cancer. And that explains the seeming difference between the women with ovarian cancer and the healthy group: The samples from the women with ovarian cancer had been run on one day and the samples from the comparison group on another. Apparently there was some subtle difference in how the mass spectrometer operated from one day to the next. The "ovarian cancer" test was really measuring nothing more than spurious signals from the machine. This is a classic example of the "batch effect," in which apparent biological differences actually stem from nothing more than batch-to-batch variation in data collection and analysis.

The *Lancet*, which had published the original article, wasn't interested in publishing Baggerly's analysis, he told me. "We yelled about this and eventually made a bit of a stink," but to no avail. Then, in 2004, Baggerly was at a convention of cancer doctors when he came across pharmaceutical salesmen from a company called Correlogic Systems, promoting the OvaCheck blood test based on this dubious technology. Baggerly had had enough. The medical journal wouldn't publish Baggerly's analysis, but the *New York Times* would. And soon the FDA stepped in and told the company to stop marketing its product until it could prove that it worked. Correlogic kept trying but could not make a compelling case. In 2010 the company filed for bankruptcy (and the OvaCheck name is now used for a completely different type of blood test for ovarian cancer).

The batch effect is a stark reminder that, as biomedicine becomes more heavily reliant on massive data analysis, there are ever more ways to go astray. Analytical errors alone account for almost one in four irreproducible results in biomedicine, according to Leonard Freedman's estimate. A large part of the problem is that biomedical researchers are often not well trained in statistics. Worse, researchers often follow the traditional practices of their fields, even when those practices are deeply problematic. For example, biomedical research has embraced a dubious method of determining whether results are likely to be true by relying far too heavily on a gauge of significance called the p-value (more about that soon). Potential help is often not far away: major universities have biostatisticians on staff who are usually aware of the common pitfalls in experiment design and subsequent analysis, but they are not enlisted as often as they could be.

Keith Baggerly stands out among biostatisticians: he actively investigates research that he has doubts about, and he doesn't hesitate to go public with what he finds. He radiates the poise and self-assurance of a sports coach, or maybe a referee. But personal charm has its limits. "There are some people who have received e-mails from me and are apprehensive about what I'm going to do if I get their data," Baggerly said. "I can understand that reaction in light of public events," but really, he insists, he's just trying to get at the truth. At times, that means digging into other scientists' work. Naturally, that doesn't make everyone happy.

A few years ago, he placed an informal wager of sorts with a few of his colleagues at other universities. He challenged them to come up with the most egregious examples of the batch effect. The "winning" examples would be published in a journal article. It was a first stab at determining how widespread this error is in the world of biomedicine. The batch effect turns out to be common.

Baggerly had a head start in this contest because he'd already exposed the problems with the OvaCheck test. But colleagues at Johns Hopkins were not to be outdone. Their entry involved a research paper that appeared to get at the very heart of a controversial issue: one purporting to show genetic differences between Asians and Caucasians. There's a long, painful, failure-plagued history of people using biology to support prejudice, so modern studies of race and genetics meet with suspicion. The paper in question had been coauthored by a white man and an Asian woman (a married couple, as it happens), lowering the index of suspicion. Still, the evidence would need to be substantial.

The two researchers, Richard Spielman and Vivian Cheung, were prominent geneticists at the University of Pennsylvania. (Spielman died in 2009.) In a 2007 study, they examined 4,197 genes in both Caucasians and Asians. Instead of looking at whether the genes themselves were different, they asked whether some genes were more likely to be switched on in one race versus another. (Genes are the message written in DNA, but most genes are silent most of the time, the spools of DNA where they reside knotted up tight. Biology gets interesting when our cells activate certain genes to read them and carry out their instructions.) They found that among those genes, about a quarter of the total were switched on or off in one race but not the other. Their paper, "Common Genetic Variants Account for Differences in Gene Expression Among Ethnic Groups," landed with a splash when published in *Nature Genetics*.

But some scientists had their doubts. Joshua Akey, along with biostatistician Jeff Leek and colleagues at the University of Washington, had performed a similar comparison between Caucasians and Africans and found a much smaller difference in that case. Because other genetic studies show that Caucasians are more closely related to Asians than Africans, Akey didn't expect to see the dramatic effect that Spielman and Cheung had reported. So he dug deeper. These experiments were run on a type of biological chip called a microarray—essentially a chip packed full of carefully arranged dots containing DNA. These tests allow scientists to make thousands of comparisons simultaneously—in this case to measure which genes are turned on and which are turned off.

The University of Washington team tracked down the details about the microarrays used in the experiment at Penn. They discovered that the data taken from the Caucasians had mostly been produced in 2003 and 2004, while the microarrays studying Asians had been produced in 2005 and 2006. That's a red flag because microarrays vary from one manufacturing lot to the next, so results can differ from one day to the next, let alone from year to year. They then asked a basic question of all the genes on the chips (not just the ones that differed between Asians and Caucasians): Were they behaving the same in 2003–2004 as they were in 2005–2006? The answer was an emphatic no. In fact, the difference between years overwhelmed the apparent difference between races. The researchers wrote up a short analysis and sent it to *Nature Genetics*, concluding that the original findings were another instance of the batch effect.

These case studies became central examples in the research paper that Baggerly, Leek, and colleagues published in 2010, pointing out the perils of the batch effect. In that *Nature Reviews Genetics* paper, they conclude that these problems "are widespread and critical to address."

"Every single assay we looked at, we could find examples where this problem was not only large but it could lead to clinically incorrect findings," Baggerly told me. That means in many instances a patient's health could be on the line if scientists rely on findings of this sort. "And these are not avoidable problems." If you start out with data from different batches you can't correct for that in the analysis. In biology today, researchers are inevitably trying to tease out a faint message from the cacophony of data, so the tests themselves must be tuned to pick up tiny changes. That also leaves them exquisitely sensitive to small perturbations—like the small differences between microarray chips or the air temperature and humidity when a mass spectrometer is running. Baggerly now routinely checks the dates when data are collected—and if cases and controls have been processed at different times, his suspicions quickly rise. It's a simple and surprisingly powerful method for rooting out spurious results.

Senior-level biologists didn't grow up thinking about problems like this. Thirty years ago, it was a minor miracle to generate fifty points of genetic data after weeks of toil. These days biologists can generate 50 million points of data before lunch. So being on the alert for issues like batch effects requires a whole new mind-set. Biology is no longer simply a descriptive science—numbers matter more all the time. That said, there are still many older scientists "who wanted to do science but didn't like math, so they thought, 'Ah ha! Biology, no problem!'" said Keith Yamamoto, vice chancellor for research at the University of California, San Francisco. "When I was in training in molecular biology [in the 1970s], friends would say, 'If you have to resort to statistics, think of a better experiment.'" As Yamamoto surveys the reproducibility issues today, he figures that mathematical and analytical problems are even more common than are the errors caused by inappropriate animal models or contaminated cell lines. And because more and more biology now revolves around "big data," scientists need to adapt to this new reality.

Errors are a reminder that new realms provide new pitfalls, but there's a big upside when this kind of science is done right. The grandest big-data prize in biology was the sequencing of the human genome itself. The code, written in simple units labeled A, T, C, and G, runs to 3 billion characters. Buried in there are approximately 23,000 sequences that encode our genes. At first, scientists hoped that simply identifying and reading these genes would in essence give them a blueprint for human beings—or at least a list of all our parts. Decoding the genome was a true tour de force, and the information it provided is a vital part of the fabric of science today. Ordinary patients may soon have their individual genomes scanned routinely to help doctors identify their susceptibility to disease.

But the scales didn't drop from our eyes when the genome was revealed. It's giving up its secrets gradually, reluctantly. Scientists constantly scour the data looking for a specific tidbit of information. Many of these projects involve wrangling huge amounts of data to sift, compare, match, and otherwise piece together a puzzle of gargantuan proportions. When scientists toil with a few genes, that's called genetics. When they wrestle with massive amounts of data all at once, that's genomics.

And genomics is just one branch of the "-omics" world. There's proteomics, in which scientists study thousands of proteins that make up the enzymes and other components of human cells (the ovarian cancer test is an example); there's transcriptomics, which looks at whether genes are turned on or off (the study comparing Asians and Caucasians falls into this category); there's lipidomics, which studies varieties of lipids, or fat molecules, that are essential parts of us; there's metabolomics. . . . Well, you get the idea.

Alas, when scientists first dived into the -omics, they did not fully appreciate what they were getting themselves into. If you can survey thousands of genes at the same time to look for correlations between them and a given disease or other effect, it's painfully easy to get it wrong. Many correlations occur just by chance, so you will quickly generate hundreds that look real but aren't. In fact, if a test is looking for a rare event, most of the time the apparently positive findings will be false. The more correlations you look for, the more erroneous findings you will encounter. And there's no telling which they are.

Over the years breathless headlines have celebrated scientists claiming to have found a gene linked to schizophrenia, obesity, depression, heart disease—you name it. These represent thousands of small-scale efforts in which labs went hunting for genes and thought they'd caught the big one. Most were dead wrong. John Ioannidis at Stanford set out in 2011 to review the vast sea of genomics papers. He and his colleagues looked at reported genetic links for obesity, depression, osteoporosis, coronary artery disease, high blood pressure, asthma, and other common conditions. He analyzed the flood of papers from the early days of genomics. "We're talking tens of thousands of papers, and almost nothing survived" closer inspection. He says only 1.2 percent of the studies actually stood the test of time as truly positive results. The rest are what's known in the business as false positives.

The field has come a long way since then. Ioannidis was among the scientists who pushed for more rigorous analytical approaches to genomics research. The formula for success was to insist on big studies, to make careful measurements, to use stringent statistics, and to have scientists in various labs collaborate with one another—"you know, doing things right, the way they should be done," Ioannidis said. Under the best of these circumstances, several scientists go after exactly the same question in different labs. If they get the same results, that provides high confidence that they're not chasing statistical ghosts. These improved standards for genomics research have largely taken hold, Ioannidis told me. "We went from an unreliable field to a highly reliable field." He counts this as one of the great success stories in improving the reproducibility of biomedical science. Mostly. "There's still tons of research being done the old fashioned way," he lamented. He's found that 70 percent of this substandard genomics work is taking place in China. The studies are being published in English-language journals, he said, "and almost all of them are wrong."

Scientists could have avoided many of the problems in analyzing big data sets had they had a clear understanding of one key concept: statistical significance. In fact, that lack of understanding plagues many areas of biomedical research. Surprisingly, many researchers have only a poor, formulaic grasp of this critical concept. And flaws in that understanding undercut many published results, from the simplest experiment on up to the million-dollar genomics scan. The term "statistical significance" is bandied about all the time. Generally, it's the lowest hurdle that a scientist has to clear in order to publish a result in the scientific literature.

The conventional (but wrong) understanding is that a study finding reaches statistical significance if there's a 95 percent chance that it is correct and only a 5 percent chance that it is wrong. This probability is frequently associated with something called a p-value. If an experiment's p-value is less than or equal to 0.05 (that's five-hundredths, or 5 percent), scientists will declare success, and many a journal will happily publish that result. But, while this definition is widely used, it doesn't mean what many scientists think it means. A result with a p-value less than 0.05 is not in fact at least 95 percent likely to be true. And in reality it sets the bar very low.

A bit of historical context will help here. One of the great scientific minds of the twentieth century was biologist and statistician Ronald "R. A." Fisher, who developed fundamental ideas that remain at the heart of statistics today. In particular, nearly one hundred years ago he invented a simple mathematical formula, called Fisher's Exact Test, to measure the strength of an observation. This test is used—in fact, abused—widely throughout biomedicine today. Here's how it came into being.

A colleague of Fisher's, Muriel Bristol, claimed that she could tell whether milk or tea had been poured into her cup first. So, out of her sight, Fisher poured her eight cups of tea, four with the tea poured first, four with the milk first.

Bristol's challenge was to identify which was which. Fisher hypothesized that Bristol could not actually tell the difference, and to measure that he devised a simple statistical test to judge the results. Note that Fisher's test applies to a very specific circumstance. It starts with the assumption that the claim (Bristol can differentiate the cups) is false, and it measures the outcome against that expectation. It does not set a magic threshold that "proves" Bristol can tell which cups received milk first. And most importantly, the test does not predict how Bristol would perform in a second round of tests. As the story goes, she identified all eight cups correctly. She had only a 1.4 percent chance of doing that if she simply chose at random—that's a p-value of 0.014, which many biologists today would take as strong evidence that she could tell the difference. In fact, the p-value provides hints but no conclusions about whether she knew, got lucky, or cheated. It's not a measure of the truth but rather a much more limited statement about the odds that Fisher was right when he hypothesized that she would not be able to tell the difference between tea prepared in the two ways.

Fisher's idea was that when scientists perform experiments, they should use this test as a guide to gauge the strength of their findings, and the p-value was part of that. He emphatically urged them to perform their experiments many times over to see whether the results held. And he didn't establish a bright line that defines what qualifies as statistically significant. Unfortunately, most modern researchers have summarily dismissed his wise counsel. For starters, scientists have gradually come to use p-values as a shortcut that allows them to draw a bright line. The result of any one experiment is now judged statistically significant if it reaches a p-value of less than 0.05.

What's wrong with this? Plenty. In the winter of 2015, the National Academy of Sciences convened a workshop to explore how better statistical methods could reduce the problem of irreproducible science. One session was devoted to the perils of the p-value. Dennis Boos from North Carolina State University offered a simple thought experiment. Say you're a scientist with a research finding that just barely reaches that magic (and arbitrary) number of significance, a p-value equal to 0.05. It's worth noting that a large number of results in the scientific literature do come in around that number. That's because studies are often designed at the outset to reach that mark. To conserve resources and save time, researchers set up a study that's just big enough to yield a result with the magic threshold of $p < 0.05$.

Boos asked his colleagues to consider what would happen if you ran that experiment again. Unless you land exactly on $p = 0.05$ a second time, there's a fifty-fifty chance the new p-value will be higher and a fifty-fifty chance that it will be lower. In other words, there's a strong chance your second experiment will have a p-value greater than 0.05 and therefore fail the traditional test of statistical significance. The exact same experiment would be deemed insignificant. That's rather startling to contemplate, because before a scientist performed that second experiment, he or she was likely under the mistaken impression that there was only a 5 percent chance that the finding wouldn't hold up. Oops. You know the saying on Wall Street that "past performance does not predict future returns?" Well, that applies to p-values as well.

The statistical elite that had gathered in the walnut-paneled lecture room at the academy nodded in knowing agreement with Boos. "We wouldn't be where we are today if even 5 percent of people [scientists] understood this particular point," said Stanford's Steve Goodman, one of the beacons of statistical reasoning. What if scientists really wanted to have a 95 percent chance that their experiment, when run a second time, would still achieve a statistically significant result? Valen Johnson from Texas A&M University had run those numbers. He said scientists could use other, more powerful statistical methods. But those devoted to sticking with p-values should aim for a result ten times more stringent: a p-value of 0.005 rather than the traditional 0.05. That tougher standard would achieve the goal that many scientists already believe they are reaching: a finding that's 95 percent likely to remain statistically significant if a study is run again. By not doing that, he said, "we're off by a factor of ten, and this is causing non-reproducibility of scientific studies."

That's a much higher bar and would by implication invalidate a large chunk of "significant" findings in the scientific literature today. That's not to say that all those results are wrong, just that scientists and journal editors place far too much confidence in them. Johnson said, "I have received a lot of pushback about this proposal to raise the bar of statistical significance from scientists, who say, 'This is going to destroy my career, and I can't do experiments anymore. I'll never get a p-value of 0.005.'" But it's not quite as daunting as it sounds. In many cases, scientists can reach that target (if indeed the phenomenon they're studying is real) if they increase the number of people, or animals, or samples in their studies by 60 percent. In the process, they'd weed out many dubious results.

Scientific societies aren't calling for that stringent an approach to biomedical research, but at least the problem with p-values has been getting some attention. In 2016, the American Statistical Association decided things had gotten so out of hand that it convened a group of statistics experts to write a statement about the pitfalls of p-values. It says what should be obvious to scientists but clearly is not: "P-values do not measure the probability that the studied hypothesis is true, or the probability that the data were produced by random chance alone." It goes on to caution against using this statistical test as the sine qua non for analysis and underscores another key point: just because a finding is "statistically significant" doesn't make it meaningful. Plenty of statistically significant results, even if found repeatedly, indicate a trivial difference that has no consequential bearing on human health.

* * *

In 2010, Uri Simonsohn, an economist at the University of Pennsylvania, came to a similarly sobering conclusion about the pitfalls of p-values through a completely different mental route. He went with a couple of colleagues to a conference on the topic of consumer behavior. "We saw a lot of findings that we thought were hard to believe," Simonsohn told me. "We noticed that when we were confronted with a finding that was hard to believe, we were siding with our intuition instead of the science. We thought that's fundamentally wrong. You are supposed to change your belief based on the evidence. We were dismissing the evidence."

That disturbing realization got Simonsohn and his colleagues wondering how hard it would be to show that something was "true" when in fact it was not. They were stunned by the answer. "It was extremely easy to find evidence for something that was not true." With little or no conscious effort, scientists can look at their data, pull out the bits that support a hypothesis, and ignore the bits that don't. Alternately, scientists can watch as their data are being generated and, the moment they reach a point of statistical significance, stop the experiment—even though more data could easily undermine their conclusion.

In a widely read 2011 paper, Simonsohn and his colleagues described this kind of manipulation. They called it p-hacking. The idea is simply to look at your data six ways from Sunday until some correlation reaches the p-value of 0.05 or less, at which point, by the conventions of biomedical science, it becomes a "significant" result. In the years since he published that paper, Simonsohn has come to realize that p-hacking is incredibly common in all branches of science. "Everybody p-hacks to some extent," he told me. "Nobody runs an analysis once, and if it doesn't work, throws everything away. Everybody tries more things."

If p-hacking weren't trouble enough, Simonsohn points to one other pervasive problem in research: scientists run an experiment first and come up with a hypothesis that fits the data only afterward. The "Texas sharpshooter fallacy" provides a good analogy. A man wanders by a barn in Texas and is amazed to see bullet holes in the exact bull's-eyes of a series of targets. A young boy comes out and says he's the marksman. "How did you do it?" the visitor asks. "Easy. I shot the barn first and painted the targets later," the boy answers. In science, the equivalent practice is so common it has a name: HARKing, short for "hypothesizing after the results are known."

It often starts out in all innocence, when scientists confuse *exploratory* research with *confirmatory* research. This may seem like a subtle point, but it's not. Statistical tests that scientists use to differentiate true effects from random noise rest on an assumption that the scientist started with a hypothesis, designed an experiment to test that hypothesis, and is now measuring the results of that test. P-values and other statistical tools are set up explicitly for that kind of confirmatory test. But if a scientist fishes around and finds something provocative and unexpected in his or her data, the experiment silently and subtly undergoes a complete change of character. All of a sudden it's an exploratory study. It's fine to report those findings as unexpected and exciting, but it's just plain wrong to recast your results as a new hypothesis backed by evidence. The fancy statistics aren't simply inappropriate; they are misleading.

Of course exploration is the essence of science. Scientists at the lab bench slip easily back and forth between the exploratory and confirmatory modes of research. Both are vital to the enterprise. The problems come when scientists lose track of where they are in this fluid world of confirmation and exploration. They have known about the hazards of blurring this line for decades. Classic papers on the topic date from the 1950s, 1960s, and 1970s. Yet press releases, news coverage, and the scientists themselves often muddy this important distinction, which probably explains why so many "discoveries" about coffee, aspirin, vitamins, or what have you end up getting overturned when the next study comes along.

Scientists may have good intentions. For example, perhaps a researcher testing a drug noticcs that it has a positive effect on a small subset of the people who try it. No doubt that's potentially exciting. That researcher may be strongly tempted to restructure the analysis to see whether, with a new set of assumptions, the unexpected result will be significant (in the statistical as well as the practical sense). When changing your hypothesis after seeing the results, "usually you have good justification for doing what you did," Simonsohn said, "but you also could have had great justification for doing 10 different things." The fundamental problem here is that scientists who do engage in such restructuring often don't even realize how badly they are abusing the tools of analysis.

At the same time, there is a strong commercial motivation to put data in the best possible light. Researchers trying to develop a drug for the market will look for statistical methods that will be most compelling to reviewers at the FDA, who weigh new-drug approval. This recurring issue is a major factor behind drug withdrawals. One disastrous example involved the anti-arthritis drug Vioxx. A best-seller when first approved in 1999, it was eventually taken off the market when more careful analysis linked it to an increased risk of heart attack. Drug maker Merck had known about the comparatively high rate of heart attacks, which its analysis wrote off as unimportant. The company argued that the patients in the comparison group actually had a lower rate of heart disease, so Vioxx was not to blame. The company lost that argument. In fact, dozens of drugs have been removed from the market after updated analysis and experience identified unacceptable risks.

Biostatisticians can see through this kind of questionable analysis—if they can see the analysis in the first place. But published methods are often vague or deliberately kept secret. An AIDS advocacy group sued the FDA to release the background data and analysis the agency had used to approve the drug Truvada as a means of preventing HIV infection. The company, Gilead Pharmaceuticals, objected on the grounds that disclosing its analytical methods could help rival companies navigate the FDA drug-approval process. The FDA sided with the drug company. Biostatisticians reacted to that story with disbelief, since good analytical methods aren't trade secrets and shouldn't be kept secret. The judge in the case agreed.

The lesson here is that solid analysis also requires disclosure and openness. That information after all allows other scientists to understand the all-important details upon which a conclusion rests. It's a critical element of science's vaunted process of self-correction. But biomedical research is often more opaque than transparent, and that contributes to the troubles with reproducibility. Here, bad incentives and bad habits are both to blame.

Chapter Seven

SHOW YOUR WORK

BRIAN NOSEK KNOWS all about HARKing. The psychology professor at the University of Virginia realized to his chagrin that he was guilty of that bad practice himself. "When you talk to graduate students in my lab they will describe this for you. We sit down at the beginning of a semester and talk about experimental design. They go and do the study. When they come back, and we're looking at the data, the first question that I have is, why did we do the study?" He can't typically remember what hypothesis they were trying to test, so he can't determine whether the results confirm a hypothesis or explore a new one. "We do both [confirmation and exploration] all the time, but it's hard to distinguish it because we're busy. We're distracted. We're just doing lots of stuff."

After thinking about his own research practices, Nosek had an epiphany. Simply increasing transparency could go a long way toward reducing the reproducibility problems that plague biomedical research. For starters, scientists would avoid the pitfall of HARKing if they did a better job of keeping track of their ideas—especially if they documented what they were planning to do before they actually sat down to do it. Though utterly basic, this idea is not baked into the routines in Nosek's field of psychology or in biomedical research. So Nosek decided to do something about that. He started a nonprofit called the Center for Open Science, housed incongruously in the business center of the Omni Hotel in downtown Charlottesville, Virginia. His staff, mostly software developers, sit at MacBook computers hooked up to gleaming white displays. Everyone works in one big, open room and can wander over to cupboards stocked with free food. The main project at the center is a data repository called the Open Science Framework.

Nosek put this new system of organization and transparency to the test in 2012 by trying to reproduce some of the studies in his field. After he floated this idea on a listserv, more than two hundred other scientists from around the world said they wanted to get in on the action. Over the next few years, this loose affiliation of psychologists selected one hundred research papers and set about redoing them. The results made news around the world. "Psychology's Fears Confirmed: Rechecked Studies Don't Hold Up," read the page-one *New York Times* headline on August 28, 2015. Two-thirds of the reproduced results were so weak that they didn't reach statistical significance. Many of those at least leaned in the same direction as the original study but could not on their own be considered evidence of an effect. About a third of the studies actually suggested there was either no effect whatsoever or even an effect opposite to what the original paper reported. There has been some pushback against these findings, but the broad conclusions still stand—and Nosek happily pointed out that his critics had made their case by accessing his readily available working material, which in itself was a triumph for transparency.

Nosek suspects that a lot of the problems he identified had arisen because the scientists running these initial studies hadn't clearly distinguished between exploratory and confirmatory research. Many were most likely exploratory studies, but the researchers ended up using statistical tests appropriate for evaluating a predetermined hypothesis. To avoid this, the Open Science Framework invites scientists to register their hypotheses in advance so that they can later demonstrate that their studies are indeed confirmatory. This is not a new idea. The Food and Drug Administration Modernization Act of 1997 requires scientists running clinical trials on potential new drugs or devices to register their hypotheses in advance in a federal repository called ClinicalTrials.gov, set up by the National Institutes of Health (NIH) in 2000. This law has another important salutary effect: some drug companies had the habit of simply not publishing studies if the results were not favorable to the drug they were investigating. This is known as the file drawer effect because studies end up getting filed away rather than appearing in the scientific literature. That tactic is now harder to hide.

This federal registration system is far from perfect. Many scientists never follow up and report their data, despite the law. And researchers still sometimes change their goals after an experiment is well under way. But at least those critiquing their work can bring these changes to light. Ben Goldacre, a doctor and gadfly in the United Kingdom, has exposed many examples of studies that don't report the results they said they would or present unanticipated results that are in fact exploratory rather than confirmatory. He hectors journals to publish clarifications when he finds evidence of this but has met with limited success.

Robert Kaplan and Veronica Irvin at the NIH set out to see whether the law requiring scientists to declare their end points in advance really made a difference. They reviewed major studies of drugs or dietary supplements supported by the National Heart, Lung and Blood Institute between 1970 and 2012 and came up with a startling answer. Of the thirty big studies done before the law took effect, 57 percent showed that the drug or supplement being tested was beneficial. But once scientists had to announce in advance what exactly they were looking for, the success rate plummeted. Only 8 percent of the studies (two out of twenty-five) published after 2000 confirmed the preregistered hypothesis.

This doesn't prove that the early studies had changed their goals in midstream, but it's highly suggestive. Kaplan and Irvin note that in many of the twenty-five papers published after 2000, scientists still reported benefits they had not explicitly set out to look for, but they dutifully pointed out that these were unexpected findings of an exploratory nature, not data supporting a hypothesis. If the scientists hadn't been required in advance to say what they were looking for, "it is possible that the number of positive studies post-2000 would have looked very similar to the pre-2000 period," they wrote.

Clearly, biomedicine is better off with transparency. Nosek envisions extending the idea far beyond the basic starting point of registering ideas in advance. The Open Science Framework allows scientists to organize their entire experiment by providing databases for depositing every aspect of the research, from the algorithms and methods scientists used to analyze their findings to the raw data itself. Data collected by a postdoc would still be easy to find long after he or she had moved on to another lab. And collaborators could easily (and selectively) share the resources involved in their project. Eventually, in Nosek's glimmering future, it should be easy to convert the text and data in a project into a publishable article.

The challenge is to convince scientists to use this system. It seems bureaucratic and cumbersome—even people on Nosek's team told me it took them a while to adjust to it. Scientists have to be persuaded that it actually solves problems that they in fact have and doesn't just create more work for them. Nosek wants to convince scientists that the time they put in at the start of the process will actually streamline their research over the course of an entire experiment—and maybe even catch methodological problems before an experiment gets under way. He is not above paying scientists to publish preregistered experiments—he landed a $1 million grant so he could offer $1,000 to 1,000 scientists as an incentive.

Sharing data and methods, whether through the Open Science Framework or some other means, would make science more transparent and in principle more reproducible. In fact, some of the major journals make data sharing a requirement of publication. You often must give your reagents to another lab that asks for them. Federal rules have openness requirements as well, but these are rarely enforced. In truth, scientists don't reliably play by these rules, even when taxpayers fund their research as a public good.

"Sharing of reagents, as long as I've worked in the field, has been highly variable," Mark Winey at the University of California, Davis, told me. And he's been a biologist for nearly thirty years. "Some labs, you just *know* they aren't going to share something, even if they publish it in a journal that has a policy for sharing. The journals have no way to enforce it." Winey said a scientist once refused to share an ingredient with him, even though a journal's policy required it. Winey complained to the journal and asked it to insist. The attempt went nowhere. The journal was "initially unresponsive, and then ineffective," he told me.

Labs gradually develop reputations. Some will send nothing; others will share everything. Winey said he's sometimes not thrilled to share if it looks like another lab is trying to "scoop" him, but that comes with the territory. Most of the time, he said, he likes to share because often he's done with a project and it's gratifying to see someone interested in carrying his ideas forward.

Sharing data can accelerate progress in biomedical research by helping researchers discover errors more quickly. During the sequencing of the human genome, scientists working on the federal effort deposited all their sequence data into a public database as they went along. And in 2001, when it came time to announce a preliminary sequence, Francis Collins, who led the genome sequencing project, and his colleagues made their announcement in the journal *Nature*. The paper has been cited more than 18,000 times, making it a blockbuster. Of course *Nature* wasn't about to publish the 3 billion or so letters that represent the human genome; instead, the article singled out a dozen or so takeaway messages that scientists gleaned from their first, excited look at the entire genome. Among the big surprises: scientists had identified 223 genes that had apparently jumped from bacteria straight into the human genome. These genes were distinct from the many that humans share with our closer evolutionary relatives, so the researchers concluded that the bacterial genes were surprising latecomers.

"I didn't believe it," said Steven Salzberg, who had a lot of experience analyzing genomes at The Institute for Genomic

Research (TIGR) (founded by J. Craig Venter, a big-time rival of Collins's group). "It's very implausible that bacteria can transfer their DNA to us," Salzberg said. Microbes would have to infect an egg or sperm cells to incorporate genetic material in a way that could be passed down through generations. "No bacteria do that. Retroviruses do that, but not bacteria." Since the genome data were all readily available, Salzberg and his colleagues had no trouble running their own analysis. He said they started working on it the very day the *Nature* paper came out. Three months later, Salzberg and his colleagues got *Science* magazine to publish their paper announcing that the purported "bacterial genes" were nothing of the sort.

Francis Collins remembers the mad scramble to get that genome paper put together. The team leaders had asked one scientist to look for genes that might have come into our genome from bacteria or fungi. "He thought he was seeing evidence of that kind of phenomenon," Collins told me. "And the rest of us looked at it and thought that's really intriguing, and while this is an unexpected finding, we couldn't find anything wrong with it. So it went into the paper." Because the underlying data were already available to anyone who knew enough to check out the claim, it didn't take long to disprove that. "But that's fine. I think that's the way it ought to be," Collins said. "You do in fact take some chances. You make it transparent and then if somebody has a more rigorous way of looking at it," he or she can set the record straight. "At least in this case it didn't use a lot of resources, and no humans were put at risk—except for embarrassment, I suppose!"

Collins himself had advocated for open genome data back in the days when he was running the Human Genome Project. That was driven in part by his desire to prevent private companies from generating data in secret and patenting thousands of human genes. But it also had a salutary effect on genome research overall. "The guys doing the sequencing aren't the best guys to analyze the data," said Salzberg, who is now at Johns Hopkins University. "I have no doubt in the world of genomics the science is going to be better if people can look at the data."

But sharing data opens up potential career hazards. "We've been scooped a couple of times," he said. For example, he was working to assemble the genome of the loblolly pine. Competitors in Sweden were at the same time sequencing the Norway spruce. Whoever published first would be able to claim having completed the first conifer genome. Salzberg was posting his data as he went along and said his competitors were clearly watching his progress, because when his genome was getting close, the other group rushed a paper into *Nature*. Salzberg says the European work was much less complete, "but the journals don't really care. They want the headlines." When Salzberg was ready to publish, *Nature* passed on the paper, pointing out that the loblolly pine wasn't the first conifer genome to be published. So it ended up in a less prestigious journal. "If we had just sat on our data and been secretive about it, they [our competitors] probably wouldn't have published when they did because they wouldn't have known where we were. So there's this incentive to not share."

This is another example of the perverse incentives in biomedical research. What's best for moving science forward isn't necessarily best for a researcher's career. "The reason we publish our results is so others can build on them. So why wait? It seems intuitively obvious; the sooner people see your results, the faster the field will move." But Salzberg ran headlong into heavy resistance to this idea about a decade ago, when he and a top scientist at the NIH were working to create a big collection of influenza genomes. At the time only half a dozen flu viruses had been sequenced, but flu researchers knew they would understand the disease a lot better if they could catalog many examples of this rapidly mutating virus. Salzberg says he was stunned to discover that many leading flu researchers wouldn't send in samples for sequencing, even if they were paid for their trouble and promised coauthorships on any paper published. "Many of the leading flu people said, 'No thank you. We don't want to share our samples.' They wanted to get NIH money to sequence their flu samples, but they wanted to sequence them and sit on them" instead of depositing them in the public database. Salzberg says some still pursue that close-to-the-vest strategy today. Eventually, enough flu researchers agreed to cooperate with the central flu-sequencing effort that it now contains tens of thousands of digital examples for scientists to use.

Researchers studying people in clinical trials also often hoard that data. Sharing isn't straightforward—scientists have to be careful not to reveal private personal information in the process, and it's not as easy to strip out every potentially revealing detail as you might think. That becomes a convenient excuse not to even make the effort. "The group that's doing the study is quite happy with that," Salzberg told me, "but that doesn't really help public health, doesn't help our understanding of cancer or other diseases." And subjects aren't in favor of keeping their data locked up in the files of the doctor doing the study. "If you ask the patient is it OK to share your data with every scientist who's working on your type of cancer, of course they'll say yes. That's why they're doing it. But they [researchers] don't ask that question! I'd like to see that change."

So would the man with ALS, Tom Murphy (see Chapter 3). As it happens, the experimental drug he was taking seemed to have a beneficial effect on him, even though it failed overall among the people in the study. And that got him thinking. Sure, the drug isn't going to be a cure for ALS, but if scientists could figure out why it may have helped him, maybe that would provide some important insight into the disease or new ideas for drugs. But Murphy quickly discovered that the world of science is not set up to work that way. All the medical information about him is scattered. His doctor has some files; he's given blood samples to other scientists for other reasons; and although researchers around the country are starting to sequence the DNA of patients with ALS, Murphy wasn't able to get anyone excited about sequencing his genome to see if it holds clues about his unusual response to the drug.

"There's a lot of good things going on, and I look at all this and say, 'Oh my God, why aren't these people working together?'" Murphy told me. When you consider how much time is spent testing potential drugs in animal experiments, if you could shorten that step, "I think they could cut eight years off the process." That sense of urgency is palpable among people with ALS today, since most have only a few years of life ahead of them, barring a major breakthrough in drug development. Murphy was frustrated beyond belief to find that so many ALS researchers simply won't share their data. It's like hiding pieces of a jigsaw puzzle. Murphy had been a military contractor. "I thought defense was tough, with all the competition, but this is a really tough one, and the lack of collaboration and sharing is . . . " He paused to find the right words. "Man, they're in the dark ages."

Transparency is at the core of a major effort to measure just how much basic cancer research can be reproduced reliably. Brian Nosek paired up with a Palo Alto company called Science Exchange in an effort to replicate fifty widely cited findings. The Reproducibility Project: Cancer Biology, as it is named, not only turned out to be a lesson in how to conduct science with maximum transparency but revealed just how challenging—and controversial—it can be to design credible experiments to reproduce the work of other labs. (It also revealed that science costs a lot of money: the team busted its multi-million-dollar budget and so had to drop about one-third of the experiments it had planned to reproduce.)

Unlike Glenn Begley, Nosek and his colleagues didn't simply handpick experiments to reproduce. They used an algorithm to identify studies published between 2010 and 2012 that had garnered a lot of attention, having either been cited in many papers or been frequently downloaded from or viewed on websites. Science Exchange and the Center for Open Science didn't have the resources to reproduce an entire paper, so they selected one or a few key experiments from each publication to run. Here's where the transparency part of the project kicked in. Once the potential experiments had been selected, Nosek's group published a "registered report" laying out their proposed experiment before actually hiring an independent lab to run the tests. This preregistration allowed outside scientists to judge the validity of the experimental design in advance. It also answered the biggest criticism lodged against Glenn Begley's study, which was shrouded in secrecy. "By making the project transparent I think that allows the community to actually look at itself a little bit more than just saying, 'Hey we've got a problem,'" said Tim Errington, who coordinated many of the replication studies at the Center for Open Science. Real numbers could flow from the result.

Glenn Begley pooh-poohed the cancer Reproducibility Project, complaining that some of the studies being replicated were so poorly designed in the first place that simply achieving the same results would be meaningless. He resigned from an advisory committee at Science Exchange in disgust. In fact, scientists at the Reproducibility Project had fretted about that same issue, as did the reviewers at the journal that agreed to publish all the registered reports and the results of the experiments that followed. *eLife* editor Randy Schekman, a Nobel Prize–winning scientist at the University of California, Berkeley, and the Howard Hughes Medical Institute, told me the reviewers made sure that the proposed experiments were scientifically valid by expanding the number of animals or adding more control groups as needed.

Errington said designing the experiments was sometimes a challenge because researchers didn't want to alter the original experiment so much that they would no longer be reproducing it. Sometimes the editors of *eLife* suggested that they study different end points, because those would have greater scientific validity. Errington explained why that would amount to a new experiment, not a replication.

Scientists whose work was being reproduced had a decidedly mixed reaction. "We had some [instances] where we could get essentially bare-bones information . . . versus some others who were incredibly engaged," Errington said. Some of the original authors spent many hours helping explain their work (notably, not a single experiment was described well enough in the original publication to be redone by anybody else). The original scientists reviewed the registered reports before *eLife* published them. Some spent a lot of time preparing fresh batches of cells or antibodies and "digging through their lab notebooks to give us raw data." Other scientists refused to cooperate at all.

"I'm dumbfounded by the silliness and the naïveté of this project," Robert Weinberg at the Massachusetts Institute of Technology told me. "These are words I choose carefully." Weinberg is a leading cancer researcher known for his science as well as his strong opinions. He was senior author of one of the fifty papers chosen for replication. He viewed the entire effort as a massive waste of time. First and foremost, he said, it often takes months for a new researcher to learn the techniques his lab uses. So Weinberg argued a one-off attempt to reproduce his work would be doomed. When the Reproducibility Project asked for more details about the experiment, he instead offered to host someone in his lab for a month to learn the techniques firsthand. "They weren't interested." Weinberg also argued that, while his particular experiment hadn't been replicated directly, other labs had observed its essential conclusions. That's always an important step in biomedical research because it shows that a result doesn't simply apply to a single cancer in a single breed of mice. On the other hand, it's not necessarily evidence that a finding is solid, as the story of transdifferentiation illustrated so clearly.

Weinberg was also skeptical about the labs that would reproduce the experiments. They would either be private labs, which drug companies often hire to carry out specific experiments, or they might be a "core facility" at a university—for example, a centralized mouse lab that cares for the animals and runs experiments for academic scientists. Though Weinberg may not hold them in high regard, these labs often have to meet higher standards than university labs do because their results can end up in drug applications that get scrutinized by the Food and Drug Administration. That didn't sway him. "What motivation do these contract research labs have to try to reproduce [my experiment]?" Weinberg wondered. "They are just being paid. And if they fail maybe they want to fail. Maybe they're not interested in vindicating me or validating the work." He said if he were to cooperate, he'd have to lay out twenty or thirty different conditions to explain the entire protocol—everything from saying when the cancer cells were last unfrozen to specifying the percentage of carbon dioxide they had in the incubator.

That was actually a point that the cancer Reproducibility Project was trying to make. Those all-important details are rarely shared. Errington says one paper gave essentially no description at all of the methods it employed. And if any tiny detail can derail an experiment, just how robust is the result? Nobody cares about an experiment that works only in a particular strain of mice with a single strain of cancer cells if it has no broader relevance. These experiments are, after all, supposed to reveal something basic about biology and, with luck, human cancer. An experiment that requires conditions so exquisite that only the lab where it originated can repeat it hardly qualifies as reproducible. That said, "not reproduced" does not always mean "wrong."

Errington said it's quite understandable that Weinberg wanted to train someone to do the experiment in his lab. "I think it's a good thing. It happens a lot in research for a good reason. But it doesn't always have to happen," and it's important that the cancer reproducibility project be consistent in how it runs these studies. If they put someone in Weinberg's lab, they'd have to do the same for every other experiment. "Imagine how much more funding we would need," Errington said. "And we would be asking a slightly different question."

In the end, Weinberg's experiment was one of the studies dropped from the project due to lack of funding. But Schekman, the editor of *eLife*, told me Weinberg's protestations were misguided. "If we don't police ourselves, you can be sure the government is going to police us, since the money comes from the government," he said. He cringed at the thought of Congress trying to ride herd on biomedical research. "Weinberg can stand on his throne and say this, but I think politically it was expedient" for scientists to take on the reproducibility project. He compared it with the 1975 conference in Asilomar, California, at which scientists voluntarily wrote rules to govern the early days of genetic engineering research.

Well aware of the trepidation that the Reproducibility Project has generated, Errington and Nosek are careful to say that the results will neither validate nor invalidate any of the papers included in their sample. After all, they are trying to reproduce just a few of many experiments in each paper. Those that succeed don't vindicate the entire paper; those that fail don't doom it. It is also likely that some experiments will fail just by chance. Here's how Errington put it to me: each experiment is nothing more than a single data point, so focusing on a single outcome would be like drawing a conclusion from a single observation. That would be bad science. Errington says the meaning will come from looking at that experience as a whole.

"We did deflate some expectations because we don't want anyone to think [our project] will do more than it really can," Errington told me. In the end, the goal is to hold up a "mirror to ourselves as a community and say, 'This is what we look like. These are our publishing practices; these are our research practices.'" If published reports are so thin on detail that they can't be reproduced, that's worth noting. And when scientists can't share their ingredients because they've already thrown them away (along with the raw data), that's worth noting as well. Errington hopes researchers will also find patterns that point out the best ways to do cancer research, so it can be more readily tested and reproduced. Ideally, the findings will be a beacon for improving the rigor of biomedical research.

Brian Nosek also hopes that this project will help biomedical scientists think differently about their craft. "Researchers care about openness. They care about reproducibility. Those are part of why they got into the discipline in the first place. They are not there cynically saying, 'I got into science in order to publish papers.' They got into science because they are curious. They want to figure things out."

Right now, the structure of science makes it difficult for scientists to live by the values that often motivated them to go into research. This is a first step. Nosek has grand ambitions: he wants to change the entire culture of science. The question is how much he can accomplish with a few well-placed nudges, "walking people down that path without even needing to think that they are going down that path . . . and then making it easy to transition to a more open and more transparent and more reproducible workflow."

But real change will also require a fresh attitude about openness. In a few biomedical fields, data sharing is now the norm. People who deduce the structure of proteins by studying X-rays of protein crystals, for example, work in a field where data, as well as analytical methods, are archived so that other labs can repeat the analysis. Young scientists brought up in the world of open-source software may be more receptive to these ideas. But it's not baked into the culture of most biomedical research.

"I've had this argument with lots of other scientists. It's hard to convince them," Steven Salzberg told me. "I'm basically saying, 'Yeah, you're putting yourself at a little bit of risk by doing this.' And they're like, 'But I don't have to do that. Why should I do that?' And I say, 'Why did you go into science in the first place? Didn't you go into science because you wanted to make the world a better place?' Yeah, they did, but that was when they were a grad student or an undergraduate. They've forgotten that long ago. They're in the rat race now." They need to get their next grant, publish their next paper, and receive credit for everything they do. They have stepped into a world where career motivations discourage best scientific practices. As a result, the practice of science has drifted far from its intellectual roots.

Across the Johns Hopkins medical school campus from Salzberg's office, Arturo Casadevall is trying to address not just transparency but other underlying issues that are hampering biomedical research. "The problems of reproducibility begin in the way we train scientists," he told me. In fact, he took the job chairing the molecular microbiology and immunology department at the Johns Hopkins Bloomberg School of Public Health because he thinks this is where he can start to reform scientific education. He wants young scientists to think more sensibly about statistics, study design, and other fundamentals of biomedical research. "There is relatively little thought in biology," he said matter-of-factly. "When people run into a problem what they do is throw more experiments at it. There is not a lot of quiet reflection about what is going on." Casadevall said he wants to put philosophy back into the PhD. Scientists need to be taught to *think*, he says. "We clearly can teach scientific thinking. But we don't do it in a formalized way."

The first step should be to teach scientists how to design experiments properly. That's usually missing from the curriculum. Graduate schools "mostly teach *facts* the first year," said Jon Lorsch, director of the National Institute of General Medical Sciences at the NIH. "They should teach *methods*." A few years ago, the NIH put out a request to the nation's graduate schools, asking for a list of the classes that teach biomedical methods. The idea was to take the best of these classes and make the curriculum more broadly available. Lorsch said the survey was a bust. Universities apparently don't offer a deep curriculum in research methodology for biomedical students.

Casadevall is planning to find ways to change those educational standards. "You walk down the hall here, and you stop graduate students, and you say how many times do you do an experiment, and they'll say, 'Three times.' You'll say, 'Why three times?'" and they'll say that's how other people in their lab do it. But Casadevall says that's all wrong. There's a rigorous way to figure out how many times you need to do an experiment to get a meaningful result that scientists can and should calculate. It takes a bit of legwork, but the results are more likely to be solid. "But that is not the way most scientists are trained. Most scientists are not trained today on the basics of epistemology or logic. . . . We need to go back to work on the basics." And he's not talking about narrow mechanics—in fact, Casadevall argues that scientists need to spend more time thinking broadly about science and less about the specifics of their discipline. That creates intellectual ruts, which, among other things, make scientists cling more stubbornly to ideas that could be wrong.

"One of the things that is maddening to me is your typical scientist is born in an area and dies in the area." Once people develop an expertise, "it's very hard to get out of a field." It becomes their social unit. Colleagues in the field are "the ones who determine your funding. They are your friends. They are at the conferences you go to," he said. "But we should be encouraging people to move and not punish them. We punish them tremendously." If you leave a field to change subjects, "they say you are not serious," Casadevall said. And the new field will also consider you fickle.

That's unfortunate: switching fields can help break ideas that are accepted as dogma. "When a newcomer comes in the first thing they usually do is they disturb the dogma," he said. They may have trouble getting their ideas published and are otherwise harassed, "but those people are incredibly important. Because they come in and they unsettle the table. It's the only way to move forward." Academia used to encourage that by granting academic scientists sabbaticals once every seven years. That's still an option, on paper at least, but for many in biomedical research, "that doesn't work anymore because everyone's writing grants and everyone's too stressed out." It's just too risky to leave your lab for an academic year, given the struggle to fund labs these days.

Casadevall himself has managed a sweeping career in science, publishing on topics from immunology to genetically modified flu viruses. He's also studied systemic problems within biology, including those that underpin the issues with reproducibility. And he's even attracted to the job of managing a department, a commitment many scientists avoid because that means less time at the lab bench. Casadevall cuts against that grain. In his view, the culture of biomedical research is badly damaged. That big problem is more important to him than the many smaller problems that individual researchers take on, hoping to make the world a better place. "If I can figure out some way of making the enterprise work 1 percent better, that would be far more important than anything I can contribute in the lab." That thinking is also starting to motivate other scientists, who recognize that the problems extend far beyond scientific training. They are starting to think about ways to fix a much deeper problem. Biomedicine's entire culture is in need of serious repair.

Chapter Eight

A BROKEN CULTURE

BIOMEDICAL SCIENCE was not always the hypercompetitive rat race that it has become in recent years. Consider the story of Charles Darwin, as he developed his theory of evolution through natural selection. That discovery became the organizing principle of biology. And the story of how it arose bears almost no resemblance to the way biology and medicine advance today. Darwin spent decades gathering observations and gathering his thoughts. He studied odd little finches in the Galapagos Islands. He pored over collections of insects. Barnacles held his interest for nine years. He spent decades breeding pigeons and soaking seeds in saltwater to see if they could survive long ocean voyages and take root across the sea. He didn't start out with a coherent hypothesis; he was simply driven by curiosity. In fact, today's science institutions would reject his approach, Arturo Casadevall told me. "He didn't stick to one thing. He had no mechanism. And yet he was able to synthesize something that is really the only coherent thing that holds biology together."

Darwin's nineteenth-century career is also different in another important way. As a gentleman-scientist, he had no need to hustle for money. And he was in no hurry to publish his discoveries. He did so reluctantly only after becoming aware that a young rival named Alfred Russel Wallace was developing a similar theory. Darwin resisted his friends' entreaties to put his ideas on paper to claim them for himself. "I rather hate the idea of writing for priority," Darwin said in a letter to colleague Charles Lyell, "yet I certainly should be vexed if any one were to publish my doctrines before me." But then, in those days of the gentleman-scientist, sailing ships, and handwritten correspondence, the stakes were mostly personal pride.

How science has changed. In contrast to the languid years of research during Darwin's day, the high pressure of competition can tempt even the best scientists into dangerous territory. Carol Greider, who shared the Nobel Prize for her discovery of telomerase, tells a cautionary tale about her early career. Her discovery triggered a race to find out more about this vital enzyme. Telomerase turns out to be composed of genetic material (RNA) and a protein component, coupled together. Greider was working feverishly with a postdoctoral researcher at the Cold Spring Harbor Laboratory on Long Island to identify the protein, while another team was in hot pursuit at the University of Colorado.

Greider and her postdoc isolated two molecules that appeared to fit the bill. Hearing the hoofbeats of competition, she rushed that finding into print. At a meeting a while later, she ran into her chief competitor, Joachim Lingner, who congratulated her but added that he was not giving up on his independent search for the telomerase protein. Greider told me she welcomed that. After all, science is based on the idea that other investigators should verify discoveries. Or not. Soon thereafter, Lingner and his mentor, Tom Cech, published a paper showing convincingly that they had isolated a completely different protein, which was in fact the actual component of the telomerase enzyme. They dubbed it TERT. "It was very clear that he was right," Greider said. She wrote another paper declaring that her proteins were not in fact part of telomerase.

"It was a pressure-to-publish situation," she said, "and some of the experiments weren't as good as they could be, but I let myself be pushed around." Science doesn't happen in a social vacuum, and in this case the postdoc in Greider's lab needed a publication in her name to help land a job. Greider felt the squeeze. On the one hand, their findings were provocative and no doubt publishable. On the other hand, the paper itself pointed out some potentially serious shortcomings in the data. If she'd had all the time in the world, Greider would have worked to resolve those lingering questions, but she says her higher-ups were complaining that her reluctance to publish was hampering the career of her postdoc. Today she chalks the episode up to her inexperience as a young investigator. Unfortunately those career pressures persist; indeed, they are even worse today.

"If you think about the system for incentives now, it pays to be first," Veronique Kiermer, executive editor of the Public Library of Science (PLOS) journals told me. "It doesn't necessarily pay to be right. It actually pays to be sloppy and just cut corners and get there first. That's wrong. That's really wrong." This perverse incentive is warping biomedical science. To keep funding flowing, researchers often choose projects that are likely to succeed quickly over those that will provide bold and deep insights. To make matters worse, there is an enormous mismatch between the number of scientists pursuing research and the funding that's available to them. There's no objective way to know just how many scientists is the right number, but absent more funding, there are too many in the system right now. As a result, scientists are actually rewarded for conducting their research with less rigor and publishing dubious results. These pressures start to mount in the very earliest days of a scientist's career. Eager students with bright ideas and high ideals find themselves swimming against a strong tide.

Kristina Martinez didn't know she wanted to be a scientist as she was growing up in a small town in rural Virginia, where her extended family raised sheep and cattle and tapped maple trees to make syrup. But, having been one of thirty-five kids in her high school graduating class, she decided to venture out into the world and give the University of North Carolina, Greensboro, a try. Martinez started studying nutrition and gradually got drawn into a laboratory where the professor was studying the biochemistry of obesity. She was hooked. She stayed on to get a PhD, as well as a degree as a registered dietician, and in 2012 set out blithely to start a career in research. "I did not know what I was getting into," she told me.

Once young biomedical scientists finish their PhDs, they go into a twilight world of academia: postdoctoral research. This is nominally additional training, but in fact postdocs form a cheap labor pool that does the lion's share of the day-to-day research in academic labs. Nobody tracks how many postdocs are in biomedicine, but the most common estimate is that there are at least 40,000 at any given point. They often work for five years in these jobs, which, despite heavy time demands, usually pay less than $50,000 a year—a rather modest salary for someone with an advanced degree and quite possibly piles of student debt.

All this would be worth the sacrifice if a research job were waiting at the end of the process. But the job market in academic research is bad and has been getting far worse. A study by the National Institutes of Health (NIH) looking at data from 2008 showed that only about 21 percent of postdocs ended up getting a tenure-track job, and the trend has been sharply downward, as the number of postdocs has ballooned. Martinez, like many of her peers, is holding onto the slim hope that she will be one of the fortunate few to land that kind of job. She had no idea how much of a long shot getting an academic research position would be when she took a postdoc position at the University of Chicago. "It just makes it scary. Now I'm in it, there's nothing really I can do about it. All I can do is the best that I can, and just hope for the best. I am trying to keep a level head about it. I've had my heart set on research so long now that I don't want to consider other options."

When she arrived at her postdoc job, Martinez started chipping away at several research projects at the same time. She also devoted her attention to helping young students, who juggle many different projects in her boss's lab. Three years into her postdoc, she had lots of stimulating ideas but no polished results to publish in the scientific literature. "Without a publication as a postdoc, you're kind of dead in the water," she said. Her boss was willing to help her apply for a federal grant so she could get funding of her own, but the University of Chicago wouldn't even consider that until Martinez had a journal article showing the results of her work. And not just any journal would do. Martinez figured she would need to get into a journal with a high "impact factor," a measurement invented for commercial purposes: the rating helps journals sell ads and subscriptions. But these days it's often used as a surrogate to suggest the quality of the research. Journals with higher impact factors publish papers that are cited more often and therefore commonly presumed to have more significance. At the top of the heap, the journal *Nature* has an impact factor over 40; *Cell* and *Science* have impact factors over 30.

These journals attract the flashiest work—though not necessarily the most careful or the most important. Martinez says her research isn't eye-popping enough to end up in one of the big-three journals. "As a postdoc my expectation for myself is to get something published in a journal with [an impact factor of] nine or above. I'd be happy with that. Fourteen would be nice. . . . That's what I'm going for." She actually prefers journals with lower profiles—the reviews are more careful, she says, and the work they publish is more detailed and nuanced. Peers in her niche field are also more likely to read them. But she put weight on the impact factor "because that's what's expected and what I need to do to push my career along."

Often times, hiring committees won't even look twice at an application if the job seeker isn't the lead author of at least one paper in a top-tier journal. Carol Greider at Johns Hopkins University said it's a poor measure of talent, but universities face a tough job in a market glutted with job seekers. "We just hired a new assistant professor in the department, and we had four hundred applications for one job," Greider said. "How do you filter those people? A lot of times the committees just scan down and look at how many high-profile papers there are." Only after winnowing the pile of resumes do hiring committees start to examine the actual research that the applicants have performed.

Journal publications have overwhelmingly become the yardstick of talent in biomedical science. Job seekers depend on them. So do scientists seeking promotion, tenure, and federal grants. "I can't tell you the number of times I've sat in a review panel and someone says, so-and-so published two papers in *Cell*, two in *Nature* and one in *Science*," Gregory Petsko, a professor at the Weill Cornell Medical College, told a crowd of postdocs at a meeting in Chicago. In those sessions, "I've raised my hand and asked in my best meek voice . . . 'Can you tell me what's in those papers?' Most of the time they can't. They haven't had time to read those papers. So they're using *where* someone publishes as a proxy for the quality of *what* they published. I'm sorry. That's wrong." Raising his voice, he continued, "A lot of great science gets published in [less flashy] journals, while crap gets published in the single-word journals." He coyly avoided naming *Science*, *Nature*, and *Cell* (he called it "Hell") and wouldn't even utter the phrase "impact factor" because he found the very concept so odious.

Veronique Kiermer served as executive editor of *Nature* and its allied journals from 2010 to 2015, when this issue came to a boil. She, too, says she's unhappy that hiring committees and tenure review boards look first at where material has been published. She's dismayed that the editors at *Nature* are essentially determining scientists' fates when choosing which studies to publish. Editors "are looking for things that seem particularly interesting. They often get it right, and they often get it wrong. But that's what it is. It's a subjective judgment," she told me. "The scientific community outsources to them the power that they haven't asked for and shouldn't really have." Impact factor may gauge the overall stature of a journal, "but the fact that it has increasingly been used as a reflection of the quality of a single paper in the journal is wrong. It's incredibly wrong."

On the December day in 2013 when Sweden's King Carl XVI Gustaf awarded him the Nobel Prize in Physiology or Medicine, Randy Schekman seized his moment in the public spotlight to publish an op-ed piece decrying the tyranny of the impact factor and, in particular, the journals *Cell*, *Nature*, and *Science*. (These journals are so deeply embedded in the everyday lives of scientists that Schekman himself had a framed cover of *Cell* in his office at the University of California, Berkeley, the issue containing one of his most celebrated publications. He winced a bit when I asked him about it and said maybe he should take it down.) If this is such a poor measure of scientific performance, I asked him, why don't universities just ignore it? "Because it's a very easy surrogate," he replied. "It's a number. Deans are bean counters. They like a simple number."

Schekman said the problem with impact factors is not only that they warp science's career system. "It's hand in hand with the issue of reproducibility because people know what it takes to get their paper into one of these journals, and they will bend the truth to make it fit because their career is on the line." Scientists can be tempted to pick out the best-looking data and downplay the rest, but that can distort or even invalidate results. "I don't want to impugn their integrity, but cherry picking is just too easy," he said. And bad as it is in the United States, Schekman said, it's even worse in Asia, "where the [impact factor] number is sacred. In China it's everything." Schekman serves on a committee in Korea that rates top-level biomedical science proposals. The scientists list as their personal goals to publish a certain number of papers in journals with high impact factors. "It doesn't matter what they're publishing," he said. The journal is all that counts. Chinese scientists get cash bonuses for publishing in *Science*, *Nature*, or *Cell*, and Schekman said they sell coauthorships for cash. That practice would fail the test of scientific integrity in the United States. Schekman helped establish *eLife* in part to combat the tyranny of impact factors. He said he told people at Thomson Reuters, the company that generates the rating, that he didn't want one. They calculated one anyway.

Sometimes gaming the publication system can be as easy as skipping a particular experiment. Olaf Andersen, a journal editor and professor at Weill Cornell Medical College, has seen this type of omission. "You have a story that looks very good. You've not done anything wrong. But you know the system better than anybody, and you know that there's an experiment that's going to, with a yes or no, tell you whether you're right or wrong." Andersen told me. "Some people are not willing to do that experiment." A journal can crank up the pressure even more by telling scientists that it will likely accept their paper if they can conduct one more experiment backing up their findings. Just think of the incentive that creates to produce exactly what you're looking for. "That is dangerous," Kiermer said. "That is really scary."

Something like that apparently happened in a celebrated case of scientific misconduct in 2014. Researchers in Japan claimed to have developed an easy technique for producing extraordinarily useful stem cells. A simple stress, like giving cells an acid bath or squeezing them through a tiny glass pipe, could reprogram them to become amazingly versatile. The paper was reportedly rejected by *Science*, *Nature*, and *Cell*. Undaunted, the researchers modified it and then re-submitted to *Nature*, which published it. *Nature* won't say what changes the authors had made to enable it to pass muster on a second review, but the paper didn't stand the test of time. Labs around the world tried and failed to re-produce the work (and ultimately suggested how the original researchers may have been fooled into believing that they had a genuine effect). RIKEN, the Japanese research lab, retracted the paper and found the first author guilty of scientific misconduct. Her respected professor committed suicide as the story unfolded in the public spotlight.

Outright fraud also creeps into science, just as in any other human endeavor. Scientists concerned about reproducibility broadly agree that fraud is not a major factor, but it does sit at the end of a spectrum of problems confronting bio-medicine. The website of the thinly staffed federal Office of

Research Integrity, which identifies about a dozen cases of scientific misconduct a year, catalogues the agency's formal findings on its website. A scroll down this page will introduce you to a former graduate student who, while working at the Albert Einstein College of Medicine, falsified data used in three journal publications and four meeting presentations. Investigators said she falsified dozens of image panels and fabricated numbers used in graphs and illustrations. An associate professor at Rowan University School of Osteopathic Medicine intentionally fabricated data leading to eight published papers and an NIH grant application. Investigators found that he "duplicated images, or trimmed and/or manipulated blot images from unrelated sources to obscure their origin, and relabeled them to represent different experimental results."

Few of these stories ever make the news. And punishment is generally mild: frequently scientists agree to work under close supervision or are barred from getting federal research grants for a few years. Many are foreign scientists who vanish from the US research scene. The Office of Research Integrity lacks the staff to investigate many cases, so its modest output is a poor measure of scientific misconduct in the United States.

Another way to measure misconduct, as well as less serious offenses, is to watch for retractions in the scientific literature. Ivan Oransky and Adam Marcus started doing that as a hobby in 2010 on a blog they set up called Retraction Watch. Oransky figured they'd post a couple of items a month. Shortly after the blog started out, "Adam was quoted saying . . . 'Our mothers will read it, and that will be fun,'" he said. But this did not turn out to be a sleepy enterprise. Retraction Watch appeared in the midst of a dramatic surge in the number of retractions. While there had been about forty retractions in 2001, Oransky said there were four hundred in 2010 and five or six hundred annually in the years since. The hobby swelled to a full-scale project, with staff and supporting grants.

Retraction Watch has fed a growing curiosity—and concern—about dubious research findings. Blog reporters chase down each new report of a retraction and try to get the backstory. Oransky and Marcus also maintain the Retraction Watch leaderboard, listing the scientists with the most retractions. Japanese anesthesia researcher Yoshitaka Fujii heads the list with more than 180 retracted papers—virtually every paper he ever published. That record leaves the competition in the dust. German anesthesia researcher Joachim Boldt weighed in with about one hundred dubious publications.

Retractions aren't limited to obscure scientists in out-of-the-way institutions. Robert Weinberg at the Massachusetts Institute of Technology has retracted five papers, including one with over five hundred citations. A graduate student in Weinberg's sprawling and highly competitive lab was the lead author on four of those papers. Weinberg says he called for an investigation after other members of his lab raised doubts about the student's work. Weinberg concluded that "everything was tainted," and nothing could be salvaged. "When people ask me about it I discourage them from trying to follow up on the work. That has been the one significant bump in the road I've had in terms of reproducibility."

Published retractions tend to be bland statements that some particular experiment was not reliable, but those notices often obscure the underlying reason. Arturo Casadevall at Johns Hopkins University and colleague Ferric Fang at the University of Washington dug into retractions and discovered a more disturbing truth: 70 percent of the retractions they studied resulted from bad behavior, not simply error. They also concluded that retractions are more common in high-profile journals—where scientists are most eager to publish in order to advance their careers. "We're dealing with a real deep problem in the culture," Casadevall said, "which is leading to significant degradation of the literature." And even though retractions are on the rise, they are still rarities—only 0.02 percent of papers are retracted, Oransky estimates.

David Allison at the University of Alabama, Birmingham, and colleagues discovered just how hard it can be to get journals to set the record straight. Some scientists outright refuse to retract obviously wrong information, and journals may not insist. Allison and his colleagues sent letters to journals pointing out mistakes and asking for corrections. They were flabbergasted to find that some journals demanded payment—up to $2,100—just to publish their letter pointing out someone else's error.

It's fair to ask why David Allison should be responsible for pointing out other researchers' errors in the first place. There's a very human answer to that question: scientists, like everyone else, hate to admit they are wrong—partly out of pride and partly because an error serves as a black mark against career advancement, tenure, and funding. "If we created more of a fault-free system for admitting mistakes it would change the world," said Sean Morrison, a Howard Hughes Medical Institute investigator at the University of Texas Southwestern Medical Center. "You have to have a culture where you don't feel the sky is going to fall on your head if you come out and say that [a finding] wasn't right."

Biomedical science is nowhere near that point right now, and it's hard to see how to change that culture. Morrison said that it's unfortunately in nobody's interest to call attention to errors or misconduct—especially the latter. The scientists calling out problems worry about their own careers; universities worry about their reputations and potential lawsuits brought by the accused. And journals don't like to publish corrections, admitting errors that sharper editing and peer review could well have avoided.

The resulting system can make a search for the truth a treasure hunt through the literature, with critiques often published in different journals and not necessarily cross-referenced. This is a result of using journal publications as the currency of science, with careers built on high-profile publications and torn down by corrections and retractions. "The literature should be more a living, evolving thing rather than full of contradictions," Morrison said. But it's hard to evolve away from an academic system that counts papers and is driven by a multi-billion-dollar publishing industry.

Often, errant studies simply fade away, sunk by their own weight, rarely referenced or used as the basis for ongoing research. They're just a line on someone's publication list and one more entry in the MEDLINE database of biomedical literature, which catalogs more than 23 million papers. But when there's an error in a splashy paper or by a big-name lab, setting the record straight can be an ordeal.

In the case of the study comparing Asian and Caucasian gene expression discussed in Chapter 6, Josh Akey, Jeff Leek, and their colleagues raised questions shortly after publication of the original paper. They wrote up their critique in a letter to the journal editor. The original authors were given a chance to tell their side of the story. You could practically hear them speaking angrily through clenched teeth. First, Richard Spielman and Vivian Cheung admitted that they had not, in fact, placed the Caucasian and Asian samples randomly on each of the microarray chips they'd studied. (That would have been impossible, given that years elapsed between the experiments with Asian and Caucasian samples.) "We regret our incorrect statement that randomization was carried out and we appreciate this chance to correct the record," they wrote. But their tone then turned prickly and defensive as they asserted that the batch effect "does not imply, or even suggest, that there is 'systematic and uncorrectable bias.'"

They did not correct their conclusion that more than 1,000 genes are expressed differently in Caucasians versus Asians. Instead, they pointed to an independent study that had identified about thirty genes that were expressed differently and pointed to nine other genes in their study that differed between Caucasians and Asians. Clearly the paper did identify some racial differences, even if it couldn't back their original claim that the difference involved a substantial share (about 25 percent) of the genes they had studied.

A peer reviewer aware of the batch-effect issue would never have allowed publication of a paper with this fundamental problem in the first place. But instead of retracting the paper, Cheung and Spielman left it standing in the scientific literature. It has now been cited more than three hundred times—in many cases by scientists who take it at face value. And the attempt by Akey and colleagues to correct the record wasn't a pleasant experience.

"We had a lot of trepidation about writing that technical comment because Richard and Vivian were much more experienced, established investigators," Akey told me, noting that he and his colleagues were just a few years into their careers. "It was not clear what the risk/reward ratio would be. With that said, everybody believes science is a self-correcting process, and ultimately we felt it was important to point this out and to let other people start thinking about some of these issues in more detail." The message in their technical note circulated among biostatisticians and geneticists who analyze this kind of data, but Akey says it's not at all clear that scientists who are trying to put these findings into a biological context understand the weakness in the Spielman/Cheung paper in particular or in other studies using similar methods. And the scientists who made the mistake were not happy to have it pointed out publicly. "Vivian at the time was really, really mad at us," Akey told me. He said her attitude has softened over the years, but even so Cheung declined to discuss the episode with me.

"Most people who work in science are working as hard as they can. They are working as long as they can in terms of the hours they are putting in," said social scientist Brian Martinson. "They are often going beyond their own physical limits. And they are working as smart as they can. And so if you are doing all those things, what else can you do to get an edge, to get ahead, to be the person who crosses the finish line first? All you can do is cut corners. That's the only option left you." Martinson works at HealthPartners Institute, a nonprofit research agency in Minnesota. He has documented some of this behavior in anonymous surveys. Scientists rarely admit to outright misbehavior, but nearly a third of those he has surveyed admit to questionable practices such as dropping data that weakens a result, based on a "gut feeling," or changing the design, methodology, or results of a study in response to pressures from a funding source. (Daniele Fanelli, now at Stanford University, came to a similar conclusion in a separate study.)

One of Martinson's surveys found that 14 percent of scientists have observed serious misconduct such as fabrication or falsification, and 72 percent of scientists who responded said they were aware of less egregious behavior that falls into a category that universities label "questionable" and Martinson calls "detrimental." In fact, almost half of the scientists acknowledged that they personally had used one or more of these practices in the past three years. And though he didn't call these practices "questionable" or "detrimental" in his surveys, "I think people understand that they are admitting to something that they probably shouldn't have done." Martinson can't directly link those reports to poor reproducibility in biomedicine. Nobody has funded a study exactly on that point. "But at the same time I think there's plenty of social science theory, particularly coming out of social psychology, that tells us that if you set up a structure this way . . . it's going to lead to bad behavior."

Part of the problem boils down to an element of human nature that we develop as children and never let go of. Our notion of what's "right" and "fair" doesn't form in a vacuum. People look around and see how other people are behaving as a cue to their own behavior. If you perceive you have a fair shot, you're less likely to bend the rules. "But if you feel the principles of distributive justice have been violated, you'll say, 'Screw it. Everybody cheats; I'm going to cheat too,'" Martinson said. If scientists perceive they are being treated unfairly, "they themselves are more likely to engage in less-than-ideal behavior. It's that simple."

Scientists are smart, but that doesn't exempt them from the rules that govern human behavior.

And once scientists start cutting corners, that practice has a natural tendency to spread throughout science. Martinson pointed to a paper arguing that sloppy labs actually outcompete good labs and gain an advantage. Paul Smaldino at the University of California, Merced, and Richard McElreath at the Max Planck Institute for Evolutionary Anthropology ran a model showing that labs that use quick-and-dirty practices will propagate more quickly than careful labs. The pressures of natural selection and evolution actually favor these labs because the volume of articles is rewarded over the quality of what gets published. Scientists who adopt these rapid-fire practices are more likely to succeed and to start new "progeny" labs that adopt the same dubious practices. "We term this process *the natural selection of bad science* to indicate that it requires no conscious strategizing nor cheating on the part of researchers," Smaldino and McElreath wrote. This isn't evolution in the strict biological sense, but they argue the same general principles apply as the culture of science evolves.

A driving force encouraging that behavior is the huge imbalance between the money available for biomedical research and the demand for it among scientists, Martinson argues. "The core issues really come down to the fact that there are too many scientists competing for too few dollars, and too many postdocs competing for too few faculty positions. Everything else is symptoms of those two problems," Martinson said. This a problem not only for people seeking jobs and promotions but for scientists fighting for grant money. Thirty years ago, about one-third of all NIH research proposals received grant funding. That figure has fallen sharply to around 17 percent. Among other things, that means the scientists who run labs often spend most of their time writing grant proposals rather than running experiments. Congress inadvertently made the problem worse by showering the NIH with additional funding. The agency's budget doubled between 1998 and 2003, sparking a gold rush mentality. The amount of lab space for biomedical research increased by 50 percent, and universities created a flood of new jobs. But in 2003 the NIH budget flattened out. Spending power actually fell by more than 20 percent in the following decade, leaving empty labs and increasingly brutal competition for the shrinking pool of grant funding. The system remains far out of balance.

Compounding the problem, states have drastically curtailed financial support for universities. It's common now for campuses to get only a small fraction of their funding from the states that proudly (and deceptively) affix their names to these institutions. To cite just one example, the marquee medical school University of California, San Francisco (UCSF), gets just 3 percent of its funding from the state of California. That means researchers must raise their own funds through grant applications, and if they fail in that increasingly competitive process, they can lose their jobs. Henry Bourne, an emeritus researcher at UCSF, says that at his high-ranking medical school, the administration no longer judges its scientists by the quality of their work; the bottom line is whether they can bring in enough money. "What we have is a Darwinian winnowing: We take them if NIH gives them a grant. And we don't if they don't. And that would be fine if the NIH was giving enough grants to ensure that we weren't rejecting people who are actually very good." But they're not, he says. Universities typically take more than half of a scientist's grant to pay for overhead expenses that states used to shoulder, back in the day when they contributed significantly to their flagship universities. Labor economist Paula Stephan at Georgia State University likens it to a shopping mall: The university owns the building and charges rent; the scientists have become the tenants, spending their grant money on rent as well as research assistants and materials. If they can't keep bringing in the money, tough. They're out of business.

Success typically requires building up a reputation by publishing a lot of flashy journal articles. To get into a high-impact journal, the story has to be unexpected (perhaps simply because it's not correct) and exciting (which may or may not make it important). Psychiatrist Christiaan Vinkers and his colleagues at the University Medical Center in Utrecht, Holland, have documented a sharp rise in hype in medical journals. They found a dramatic increase in the use of "positive words" in the opening section of papers, "particularly the words 'robust,' 'novel,' 'innovative,' and 'unprecedented,' which increased in relative frequency up to 15,000%" between 1974 and 2014.

To get a paper in a top journal, scientists also need a squeaky clean story—free of peripheral observations that could raise any questions about the central findings and with no weak statistical findings. Of course, the real world of biomedicine is complex and untidy, so superclean studies actually merit suspicion rather than the public spotlight. "There is a lot of pressure for beautiful results or really clean results," said Ken Yamada, a senior researcher at the NIH and editor of an academic journal. "And I think there used to be logic to it. Many years ago if somebody showed beautiful data, that almost always implied that they had to have repeated the experiment multiple times." It strongly suggested robust results. "But nowadays if people adjust the appearance of things or just pick the single perfect example—and there are a lot of less convincing examples—there's no way for you to know because all the data aren't shown. It looks beautiful. It's convincing. Pictures don't lie"—or so we readily believe.

Yamada says this isn't necessarily a deliberate attempt to deceive. "There's a lot of misunderstanding [among scientists] about the integrity of scientific information. I personally think that part of it comes from just the removal of red eyes from photographs, making things look prettier just in everyday life." But prettifying can easily go too far. For example, scientists employing a technique called single-particle electron microscopy use computer software to help them make sharper images. Sometimes scientists feed in a mathematical "model" representing what they're expecting to see, so if something like that pops up in their field of view, the software will recognize it and make the image sharper. (Your digital camera does something like this when it stabilizes an image, adjusting the pixels that would make the picture blurry.) Maxim Shatsky and Richard Hall at the Lawrence Berkeley National Laboratory showed how this technique can lead researchers astray. They used the iconic photograph of Albert Einstein sticking out his tongue as the model they fed into a computer. The image processing software was programmed specifically to look for hints of the Einstein image and enhance any signs of it. Shatsky and Hall then fed the computer 1,000 images of nothing but static. Lo and behold, the software "correction" produced an unmistakable picture—of Einstein sticking out his tongue. Richard Henderson at the Medical Research Council Laboratory of Molecular Biology in Cambridge, United Kingdom, said he finds completely misleading images in the literature based on extreme "corrections" like this, including, in one instance, a bogus view of a vital protein component of HIV. "One must not underestimate the ingenuity of humans to invent new ways to deceive themselves," he wrote.

The deep structural and funding problems throughout biomedicine are not news to anybody involved in the enterprise. In recent years, the topic has gone from a subject of idle shoptalk to a matter of serious discussion. In fact, that's how Gregory Petsko came to be making tart comments about the journal "Hell" to young scientists in Chicago. He was speaking at a meeting put together by postdocs, including Kristina Martinez, to grapple with these existential questions. Chicago-area postdocs spent nine months in their not-so-spare time assembling a meeting under the rubric "Future of Research," patterned after similar events in San Francisco, Boston, and New York. Postdocs realize they are inheriting a mess—a situation that not only jeopardizes their careers but makes it hard for them to solve the big problems and advance medical research, as so many have dreamed of doing.

During the morning's presentation, the sixty-seven-year-old Petsko told them, "If we really care about the culture of science, it's up to the old fogies of the world to do something about it." His generation was running the show when the system broke. But these young scientists seemed determined to identify their own ways to reform the culture of biomedicine: improve the mentor-protégé relationship, find better ways to collaborate (and be rewarded for that), fight against the tyranny of journal impact factors, and avoid the pressure to exaggerate and hype results.

In 2014, some leaders of the biomedical enterprise decided it was time to start a serious conversation about these issues. Bruce Alberts (then president of the National Academy of Sciences), Marc Kirschner (chair of systems biology at Harvard), Shirley Tilghman (former president of Princeton), and Harold Varmus (then head of the National Cancer Institute) wrote a paper titled "Rescuing US Biomedical Research from Its Systemic Flaws." They acknowledged that they could no longer let these structural problems fester and called for a meeting of minds across biomedicine to seek solutions. To prime the pump, they suggested a few of their own.

The article wasn't simply a predictable lamentation about the need for more money in research. Even optimistic increases in funding won't create the needed balance. Instead, scientists and their institutions need to make some hard choices—for instance, reducing the role of postdocs and hiring more scientists into regular staff jobs. The article became the centerpiece of a wide-ranging conversation among people in the field. Four months after it was published, the authors convened a planning meeting of about thirty people, representing universities, scientists, students, and government agencies, to set an agenda for an even wider discussion. The focus was not reproducibility per se but the field's underlying pressures: the hypercompetitive environment of biomedical science. The meeting ended in discord, with no agreement even about how to approach a larger conversation. Attendees did agree, though, on one point: "Doing nothing is not an option." The four leaders didn't give up entirely: they created a small organization called Rescuing Biomedical Research to keep pushing for systematic change—including a discussion about reducing errors in science.

Those with the most economic power—the federal funding agencies—can't simply impose solutions from above. "The NIH is terrified of offending institutions," said Henry

Bourne at UCSF. Conventional politics in part drives congressional funding for biomedicine. Members of Congress support institutions in their districts because local economies grow when federal dollars flow to universities and medical centers. Congress has also funded biomedical research because so many politicians have a sick relative or a dying friend and want to support the search for treatments and cures. But Bourne fears enthusiasm for that more foresighted reasoning has waned. "The government, and actually the American people, have suddenly realized that they're spending a lot of money and cancer isn't yet cured, so to speak. We bragged that we would cure cancer, and then it turns out we didn't." Bourne worries that "everyone suddenly thinks research is terrible and it's not worth anything." He doesn't hold that view himself, naturally, but he does understand how frustration arises from the slow pace of progress.

Bourne has ideas about how to improve matters. For example, he'd like his university to establish an endowment to fund key professors' base salaries to reduce the do-or-die scramble for research dollars. But he also believes scientists themselves need to change. "I think that is what the real problem is—balancing ambition and delight," he told me. Scientists need both ambition and delight to succeed, but right now the money crunch has tilted them far too much in the direction of personal ambition. "Without curiosity, without the delight in figuring things out, you are doomed to make up stories. Occasionally they'll be right, but frequently they will be not. And the whole history of science before the experimental age is essentially that. They'd make up stories, and there wouldn't be anything to most of them. Biomedical science was confined to the four humors. You know how wonderful that was!" Hippocrates's system based on blood, yellow and black bile, and phlegm didn't exactly create a solid foundation for understanding disease. Bourne argued that if scientists don't focus on the delight of discovery, "what you have is a whole bunch of people who are just like everybody else: they want to get ahead, put food on the table, enjoy themselves. In order to do so, they feel like they have to publish papers. And they do, because they can't get any money if they don't." But papers themselves don't move science forward if they spring from flimsy ideas.

There has never been a more important time to get this right. Biology is in the throes of a shift from small studies to big data. In this new world, quality is paramount. Scientists are starting to mine massive amounts of data to discover unsuspected links between genes, behavior, biochemistry, and disease. This is the foundation of what's being called "personalized medicine" or "precision medicine." The NIH and Barack Obama's White House recognized this as a major new initiative. Indeed, it could be the future of medicine. Unfortunately, some of the foundational work has started off on less than rigorous footing. And without reliable, consistent information to work with, precision medicine may find itself facing the dreaded phenomenon computer scientists have memorably labeled "garbage in, garbage out."

Chapter Nine

THE CHALLENGE OF PRECISION MEDICINE

WHEN CAROLYN COMPTON was working as a pathologist at Massachusetts General Hospital, one of the world's most celebrated medical centers, she never knew when a cancerous colon removed in an operating room would show up at her lab. It could take days. "I can tell you that there was no urgency" to get a colon from the operating room to the pathologist who would diagnose the disease. "A big colon gets put into a bag. It sits in the operating room until a circulating nurse gets around to putting it in the holding refrigerator in the operating room. At the end of the day the same guy who delivers the mail at Mass General comes around and puts it in a cart. He takes it two buildings over, to the pathology department. There it goes onto another bench to get logged in by a technician and it goes into a refrigerator. If it's a three-day weekend, the resident on call doesn't come in until Tuesday, opens up the colon and takes a piece of the cancer and puts it into formalin," a preservative.

This delay was not a problem for the patient when Compton worked at Harvard in the 1990s, and it isn't one today. Pathologists can still stain that piece of colon, study it under the microscope, and diagnose the type and stage of cancer. "That [drawn-out process] met the standard of care, and still does," Compton told me. But she has gradually come to realize that this rather casual attention to tissue collection and preservation spells real trouble for biomedical research that might be conducted on a sample of colon cancer. These tissues are perishable, so studies that depend on fine molecular measurements are likely to be irreproducible.

These days, scientists are trying to wring a great deal more information out of the tissue than they can see through a microscope. Precision medicine could potentially correlate specific snippets of DNA, proteins, and other molecules with disease diagnosis or prognosis. Many of these molecules are quite fragile. Compton says even the anesthesia used in the operating room to knock out the patient can affect them. These molecules can change more when surgeons cut off the blood supply to the tissue to be removed. And once the organ is out of the body, the stability of those critical biological molecules will vary depending on the room temperature and—significantly—the amount of time the tissue sits around before it's preserved. "We will not have precision medicine unless we can fix this problem," Compton said.

Compton, now at Arizona State University, says it's taken quite a while for pathologists to realize just how important all those factors could be. One wake-up call came from the lab of David Hicks at the University of Rochester. Starting in 2006, he was trying to unravel a serious medical mystery. The Food and Drug Administration (FDA) had approved a test to help diagnose a particular variant of breast cancer, called HER2-positive. The test itself was certified as very accurate and reliable. Yet about 20 percent of the time it reported that a tissue sample lacked the HER2 trait, when in fact it was present; and up to 20 percent of the time the test "found" the HER2 trait even though it was not there. That's bad either way. Either women who could benefit from Herceptin, a drug that targets HER2, weren't getting it, or women were receiving an expensive drug that was not only worthless to them but also had side effects.

But if the test itself wasn't at fault, the problem must lie in its use. Hicks's colleagues solved at least part of that mystery with a simple experiment. They let breast tissue from biopsies sit out for an hour or two before testing it. And that was enough to degrade the sample and turn a positive result into a negative one. The molecule detected by the HER2 test breaks down at room temperature. "You can have the best test in the world and still get the wrong answer if you bugger up what you are testing," Compton said.

That observation sparked action. Two leading professional societies, representing pathologists and clinical oncologists, drew up new rules in 2010 that, among other improvements to reduce false readings, required preservation of breast tissue within an hour of surgery. That has helped make the HER2 test significantly more reliable. But Compton noted with exasperation that breast cancer is the only cancer for which doctors are required to pay attention to the clock after surgically removing tissue. There are no standards for treating samples from the more than two hundred other forms of cancer. She's not simply concerned about patient care. She's thinking about how to improve the reliability of scientific research based on those samples.

Compton explained that pathologists come in two varieties: clinical pathologists, who diagnose disease, and anatomic pathologists, who do research. Clinical pathologists follow long lists of federal standards and professional practices. "In laboratory medicine, accuracy of measurement and reproducibility of measurement is everything. *Everything*," she said. "In fact all regulation related to laboratory medicine is focused on your ability to calibrate and reproduce reliable analytic results from run to run from day to day from lab to lab." After all, an error can lead to misdiagnosis, with life-and-death consequences. "You would never believe a result from a [blood sample] if you hadn't handled the blood tube properly. You would just order another draw."

Anatomical pathologists and allied medical researchers, on the other hand, tend not to think as much about the quality of their starting materials. "They would come to the pathology department and say, 'Can I get twenty [paraffin-preserved] blocks of colon cancer?'" Compton said. "They were glad to get their hands on anything." They would take those starting materials back to their labs, "spend a huge amount of time and money analyzing them and then get results that nobody could interpret—and *they* never could actually interpret!" Compton says there are no national standards for handling tissue in research labs.

Medical researchers are already getting a taste of the daunting problems that will affect precision medicine if they aren't absolutely scrupulous about sample collection. Paul Tempst and colleagues at the Memorial Sloan Kettering Cancer Center in New York City were running a study involving proteins taken from blood samples. The scientists were initially encouraged to find a difference between blood samples taken from cancer patients and those taken from healthy people, but Tempst became concerned that the result might simply reflect a batch effect. Sure enough, he eventually tracked down an unexpected culprit: the test tubes in which the blood had been collected. Tempst realized that samples from the healthy patients were collected in a clinic, while samples from people with cancer came from a hospital. And it turned out that the hospital used one type of test tube to collect blood, the clinic another. That seemingly trivial difference was enough to render his results meaningless.

Collecting the starting materials correctly is an essential first step, but it's just a start. Carolyn Compton started thinking about these issues when she was working at the National Cancer Institute (NCI). She was not alone. Her friend, Anna Barker, then NCI deputy director, was growing alarmed as well. After a dinner conversation veered deeply into the topic, "that really set me on a path to say, 'Let's create some standard operating procedures and best practices,'" Barker told me. Samples were one thing, but "how do you collect the data? How do you exchange the data? How do you analyze the data?" Individual researchers invent their own ways to do this. "Everybody wants to do their own thing," Barker said. Even though that's the ethos of biomedical research, she realized that simply would not work once scientists tried to pool their data. That way lay mayhem.

With that in mind, in 2004 Barker set out to assemble The Cancer Genome Atlas (TCGA), a massive compendium that would map out a multitude of genetic changes associated with various cancers. And to make sure the data were comparable, she didn't fund individual investigators. Instead, she contracted with researchers to perform specific tasks and to meet specific standards. "It was based on creating a situation where we would get reproducible data. So that means we controlled everything about this project," she said. Barker specified how tissue would be collected, handled, stored, and sampled, how the DNA would be sequenced, and how the results would be analyzed. The work may not have felt creative to the scientists doing it, but Barker came away from the decadelong project with a sense that she had not only gathered a pile of useful data but convinced scientists to work toward one collective goal rather than pursuing individual projects.

That experience is by far the exception. Scientists are reluctant to create standards and even slower to adopt them. Something as commonsensical as authenticating cell lines has been a slog. Yet standards are hardly a new idea in science and technology. "We have tons of standards. We have more standards than you'd ever want to think about," Barker said, for everything from lightbulbs and USB ports to food purity. But they don't permeate biomedical research. "How many standards do we have in whole genome sequencing? That would be none at this point," she said. Or how about a standard to search for mutations in genome sequences? "Nobody does it the same way." Scientists have proposed many standards over the years, but their colleagues are often unaware of them, and, in any case, there's no easy way to impose them on an entire field.

Biomedical science just can't keep on going this way. "Biology has become quantitative," she said. "That's a transition that we're just beginning, and it's going to leave a lot of people behind. We haven't trained a lot of our biologists to think mathematically or to understand or analyze data. Most people understand that we're in the digital revolution. Well, your genome is digital data."

As John Ioannidis discovered, the early efforts to collect meaningful genome data were a miserable failure. The scientific literature swelled with tens of thousands of papers reporting the discovery of a genetic marker for this or that disease.

Attentive news consumers will remember many occasions in which researchers announced "a gene" for schizophrenia or colon cancer or leukemia. Of those, fewer than a dozen putative discoveries have been solid enough to lead to an FDA-approved blood test or therapy. Companies will happily screen your blood for suspicious traits, but most of the results aren't concrete enough to form the basis for medical treatment. And it has been a struggle to get reproducible results, despite many millions of dollars in taxpayer investment.

This is especially true for studies in which scientists are trying to use genomic information in the quest for new cancer drugs. Even results from the world's top laboratories sometimes disagree. Jeffrey Settleman and colleagues at Massachusetts General Hospital and the Harvard Medical School set out to look for new cancer drugs by screening compounds in more than six hundred cancer cell lines. Each line had been genetically fingerprinted, so the scientists could not only identify cancers that responded to individual drugs but look for genetic patterns as well. Drugs can be effective in different cancer types if those cancers are genetically similar. In 2012, the group published the results from more than 48,000 individual tests involving 130 potential drugs in these cancer cells. In collaboration with another top lab, the Wellcome Trust Sanger Institute in England, the work identified a few promising leads, linking drugs, genes, and specific cancers.

At the same time, a second consortium had set up a similar massive experiment. Scientists at the Broad Institute in

Cambridge, Massachusetts, joined forces with the Novartis drug company to screen twenty-four different drugs in nearly five hundred cancer cell lines. The combined efforts cost tens of millions of dollars and constitute the largest public collections of genetic and drug data in the world. The Broad team published its first findings in the same issue of *Nature* as the team from Mass General and also highlighted a few leads for future drug development.

John Quackenbush and Benjamin Haibe-Kains, at the Dana-Farber Cancer Institute in Boston, decided to compare these two efforts to see if their results matched—a potentially powerful way to validate the data, since the two research teams used the same starting materials but different testing procedures and analytical methods. "We thought, what could be better than this?" Quackenbush told me. Quackenbush and Haibe-Kains identified fifteen drugs and 471 cell lines common to both experiments. The following year, they published a bombshell in *Nature*: the results of the two experiments showed almost no correlation. Only one of the fifteen drugs really seemed to behave the same way in the two studies. "If you want to build a predictor of drug response, and you're using those data, you're in trouble," Haibe-Kains told me. Quackenbush chimed in: "How can you ever hope to take data from these cell lines and make a prediction you can take into patients? It just doesn't work."

Their paper caused quite a stir in the field, generating both consternation and some criticism of their own analysis. Quackenbush and Haibe-Kains accepted the criticism and tweaked their findings. That "moved the needle a little more on the side of consistency" between the two data sets, Haibe-Kains said. Consistency for two or three drugs improved, "but we are still extremely far from taking one data set, validating it on another data set, and jumping directly to the patient."

Two years later the authors of the original studies fired back with their own reanalysis. Using a more relaxed standard and some unorthodox statistical techniques, they concluded that the results, while far from a perfect fit, gave them a correlation that "seems reasonable" and "acceptable," especially for the large effects. More than 90 percent of the time, neither experiment showed that a drug was likely to work, and a lot of the disagreement had to do with variation in the less dramatic results, which author Levi Garraway argued weren't very useful in any event. "The reality for most cancer drugs is that most patients don't respond," he said. The rare individuals who do are the interesting cases. He said his studies focus on identifying those rare dramatic effects, and those kinds of results are more consistent across the two studies.

But Quackenbush and Haibe-Kains expected much more from this rich and expensive trove of data: they hoped to find new clues to disease by finding combinations of less dramatic results to provide medically useful insights. That would be hard to do because there wasn't even agreement about where to draw the line between uninteresting noisy data and interesting but less dramatic findings. The conflict spiraled into a heated back-and-forth over who was right and who was wrong. "We wasted our time on defending our position and point of view instead of working together to make it [the analytic process] better," Haibe-Kains said. He ended up hiring a postdoc to work full time on science related to the controversy. "We both have lost something there."

Garraway was also displeased with how events unfolded. He noted that he and Quackenbush are both affiliated with Harvard's Dana-Farber Cancer Institute, "so I will admit to being somewhat disappointed that if you publish a paper like that we would never have talked about it in advance," he told me. (Quackenbush said he had reached out to Garraway's group but was rebuffed.) That conversation seems even less likely to take place now, since Garraway didn't discuss his own reanalysis either before publishing it. Quackenbush, a computational biologist, would have panned it. He said he would never have allowed a student of his to use the reanalysis techniques that Garraway and his colleagues applied. "I'd ask them what demon of data dredging possessed them to go and make that kind of analysis."

As is often the case in science, more data at least partly resolved the dispute. Genentech scientists ran similar cell line studies and compared their results with those of the two other efforts. Their findings were quite similar to those of the Broad Institute but also agreed with some of the Massachusetts General Hospital research. But that third analysis also focused specifically on large effects and set a lower bar for "agreement" than Quackenbush and Haibe-Kains had.

Garraway said the question isn't whether the two labs would always produce the same exact results. Since they used different techniques and tests, the experiments were not designed to replicate one another exactly, the way running a second test with the same ingredients would. He said agreement between two different techniques enhances confidence in the result. When both are technically valid and they don't agree, the divergences can potentially reveal something important about cancer biology—if you can explain why the results differ.

The story doesn't end there, however. Even before this problem arose, Peter Sorger at Harvard had been troubled by a deeper question: Are any of these experiments producing rigorous findings in the first place? He had serious doubts. For decades, scientists analyzing cancer cell-line tests had been ignoring some critical biology, and that called much of the work into question. For example, these tests typically don't take into account that different types of cancer cells grow at different rates. As a result, scientists can be fooled into thinking that a drug is effective at slowing cancerous growth, when in fact the cells are just proliferating slowly to begin with. Sorger ran a series of experiments to show that the standard approach was horribly flawed. He then developed a straightforward method to correct for that.

When Genentech published its findings, they contained the right details, enabling Sorger to apply his correction to the data. And the result: only 40 percent of the correlations the company scientists had reported between gene variations and drug sensitivity remained valid after the correction was applied. In one case, the Genentech study found a thousand-fold increase in drug sensitivity among ovarian cell lines that carried a certain mutation. But when Sorger corrected for the fact that the mutation made the cells proliferate much faster to begin with, the drug effect simply disappeared.

And it's not just cell proliferation rates that matter. The density of cells in a flask can also have a profound effect on a result. Many academic researchers aren't correcting for that, either. "From a theoretical standpoint, the way in which drug response had been traditionally measured is just not sound," Sorger said. When he talks to oncologists about his findings, he said, they are aghast. "We use that information to determine how we prescribe drugs to patients," he said. "I think that makes the whole issue much more concerning."

He has devoted about one hundred people in his lab to sorting out these issues, which are fundamentally a matter of reproducibility. That's not a direction he expected to take. And though it has become a truism that research on cancer cell lines doesn't translate into meaningful treatments, Sorger believes it does not have to be so. He worries that the field could give up on the approach altogether, when researchers could instead improve it by thinking more deeply about the underlying biology and applying those lessons. And Sorger is deeply frustrated that the serious issues he has raised don't seem to have sunk in—most experiments still use the old techniques.

Sorger has been arguing that simple changes to test protocols would enable researchers to gather much more meaningful data. The problem, though, is that scientists have already sunk tens of millions of dollars into doing the work one way, and it's not an easy call to go back and do it all again with different procedures. "One of the things is that when you start a large-scale project you often have to make trade-offs," Levi Garraway told me. "You can't always get everything for every project." That said, when a powerful new technique comes along, labs do sometimes go back and redo many experiments. Garraway agreed that Sorger has made a good argument for changing the way these kinds of tests should be run. "I'm not promising it will happen, but it's eminently possible," he said.

Scientists pursuing the dream of precision medicine also have a great deal of work to do in order to make biomarkers more trustworthy. Researchers know that if they can find reliable biomarkers to diagnose and track the progression of a disease, they will learn much more quickly whether a potential drug works. For example, researchers long ago learned to measure the amount of HIV in the blood, and by doing that they could readily tell whether a drug would beat back the virus. That greatly accelerated drug development because pharmaceutical companies didn't have to wait to see if people lived longer—they could simply measure the effect of the drug on viral load. "You still have to make sure you get the dose right and you understand the toxicities and everything, but it tells you very quickly whether or not the drug is going to have an effect," said Janet Woodcock at the FDA.

But most biomarkers reported in the scientific literature have been dismal failures. Of all the problems in biomedical research, "the irreproducibility and the lack of rigor on the biomarker side is probably the most painful," Woodcock said. She blames academic researchers for insufficient rigor in their initial efforts to find biomarkers. "The biomedical research community believes if you publish a paper on a biomarker, then it's real. And most of them are wrong. They aren't predictive, or they don't add additional value. Or they're just plain old wrong." To figure out exactly what a biomarker does and doesn't tell you, "you have to do a lot of work. People don't want to do that work. So this problem isn't just in the laboratory. It extends over into the clinic in a big-time way."

Woodcock sees a lot of potential for biomarkers. For example, in her field of rheumatology, doctors have known for many years that osteoarthritis progresses rapidly in some people and very slowly in others. There must be a biological reason for that difference, and discovering that could point to new approaches to treating the disease. Osteoarthritis strikes millions of people, causing disability, joint pain, and expensive joint-replacement surgeries. "There have been like ten thousand papers published on osteoarthritis biomarkers with no rigorous correlative science going on," Woodcock said. A number of projects, funded by both government and industry, are sifting through all of those leads to find biomarkers that are actually reliable, "but the effort, compared to the magnitude of biomedical research enterprise, it's like spit or something" in an ocean of dubious data. "The dirty little secret is it costs tens of millions of dollars. It's expensive."

Drug companies have sometimes made that investment, and that's why there are a few successful tests on the market today, particularly for genetic profiling of breast cancer. Those tests show the promise of these technologies, but they are the exceptions. Part of the problem is that a successful biomarker isn't likely to be a big moneymaker, certainly not compared with a new cancer drug, which can sell for tens of thousands of dollars for a course of treatment.

Scientists are also reluctant to admit, even to themselves, that they are facing a dead end, especially if they have built an entire lab around a particular idea. Josh LaBaer at Arizona State University said that if you work for a pharmaceutical company and your research flops, the company will probably just assign you to a new project. "It's not so easy in academia," LaBaer said. "I don't know if there's a way in academia to make it so that people can retain their positions but nonetheless walk away from data that isn't looking encouraging. I think that's a big part of this reproducibility problem. There's this need to stick with what you find because your career depends upon it. If you could report it as negative and say it didn't work and still survive, I think you'd be more inclined to do that."

LaBaer has been using his position as an editor of the *Journal of Proteome Research* to stanch the flow of biomarker papers that go nowhere. "I've gotten pretty tough lately," he told me. He won't accept a paper that simply reports a correlation between a biomarker and a medical condition. He tells authors they need to use that observation to generate a testable hypothesis and then test it. He won't even review those papers anymore. "I send them back." It's not helpful to have the literature filled with papers that rarely pan out. "I've gotten a lot of heat for that, but that's really been my policy." I asked him whether scientists actually do follow up with more rigorous studies, or if they simply take their papers to one of the thousands of less selective journals to get the work published anyway. "I don't know," LaBaer shrugged. And it matters. "It's going to impact all of us, because more and more people are doing massive literature searches to build databases of information by summarizing the literature automatically." That makes it even harder to find biomarkers that could actually make a difference for diagnosing and treating disease.

Anna Barker is trying to find a way through this morass—starting with tissue collection and including drug discovery and biomarker validation—by taking a fresh approach to one of the most challenging cancers: glioblastoma. "Nothing has changed in this cancer for the last hundred years," she said, alluding to the hundreds of failed efforts to find a viable treatment. Barker organized a coordinated effort to take on this hardest of hard cancers, figuring if she can crack it, people will have to sit up and pay attention to how she did it. "We want to try to do everything right from square one," she said. That starts with highly regimented collection and testing of tissue samples, along with treatment procedures that are followed to the letter from one hospital to the next. This is an international effort, involving hospitals in Australia and China (which funnels its glioblastoma patients to four hospitals nationwide).

The study itself represents a major departure from the way clinical trials are typically done. Usually a drug company pays a group of researchers to test a single drug. A large-scale trial involves hundreds or thousands of patients, and once the plan is set, the study continues unchanged until there's either an obvious victory, an obvious problem, or all patients have been enrolled and examined. Barker's glioblastoma study, called GBM Agile, is instead an "adaptive trial," which means researchers try to learn something from every single patient as the study progresses. If one patient seems to respond to treatment better than others, scientists will look for genetic clues and other leads that could help them modify the trial for the next patient. And the same team will test an array of drugs from a variety of manufacturers, singly or in combination, depending on where the ongoing study is leading them.

Researchers studying breast cancer pioneered this concept in the early 2000s. But adaptive trial designs are seldom used, in part because they are tougher than standard clinical trials to get approved by the FDA. But Barker has navigated that challenge. She has also developed a network of about 150 collaborators working toward the project's common goal of finding a viable treatment. In this way the trial design gets away from ego-driven research and instead rewards collaboration. "There's got to be a better way," she said. "There's *got* to be a better way."

It might seem that Barker is stacking the deck against herself by choosing such an unyielding cancer as the project's focus. But she doesn't see it that way. "We need some demonstrable success in the rare tumors, because frankly most diseases are going to become rare diseases." That is the paradox of precision medicine: instead of homing in on a common treatment for each disease, researchers may end up with a much more complicated and expensive problem to confront. Each patient will have a much more specific genetic diagnosis. And doctors will no longer be treating two hundred different types of cancer; they will potentially be treating thousands of unique diseases, each fine-tuned to the genetics of the individual or to the genetic pattern of the tumor.

The deepest challenge in realizing the potential of precision medicine is in changing the underlying incentives in biomedical research. That means reengineering the culture. The question is how to do that. Step one is to make sure the problems and the perverse incentives are well understood. Step two is to figure out how to create new incentives for scientists, universities, and funding agencies. If this is starting to sound like the germ of a research initiative, it is. In fact, a whole new field is emerging—one designed to study problems in how scientific research is conducted and to identify solutions. It's called meta-research.

Chapter Ten

INVENTING A DISCIPLINE

STEVE GOODMAN has spent much of his career thinking about the many ways that medical research can go awry. It was the intellectual drumbeat that kept him stimulated as he performed more mundane tasks as a biostatistician and epidemiologist at Johns Hopkins University, such as helping scientists there design valid clinical trials. Finally, Goodman decided to turn his full attention to issues of rigor and reproducibility in biomedicine. He moved to Stanford University in 2011 and two years later cofounded a new endeavor called METRICS, an acronym for Meta-Research Innovation Center at Stanford. "We do research on research," Goodman told me. "To figure out what's wrong and how to make research better, you have to study research. That's what meta-research is. It's not like metaphysics. It's real. And we look at real things."

"I wanted to call it the Center for Medical Truth," he told me. "This was roundly nixed." That rejected name reveals the depth of Goodman's concern about the heart of medical research.

Not only does METRICS have an unusual mission; it has an unlikely history. The codirectors, Goodman and John Ioannidis, are not natural partners. Goodman is deliberative, while Ioannidis moves rapidly from project to project, publishing dozens of papers every year. The two scientists even faced off in a very public disagreement a decade ago. But their different styles and approaches may actually be an asset as they seek to disentangle the many factors responsible for the lapse of rigor and reproducibility in biomedicine.

Both scientists had spent years doing research that formed the foundation of meta-research. During his medical training in Greece and later at Harvard, Ioannidis realized that the medical literature was deeply unreliable. He launched his career studying the shortcomings of research involving human subjects. "Most of the time what we would find out was that the data were horrible," he told me. "The analysis had major problems. There were strong biases. . . . And most of the time, if you had to be honest, you would conclude despite all this data, I really don't know what's going on here." In the 1990s, he and Goodman independently became part of a wave to clean up the methods used in clinical medical research. It was a transformative time for medicine, because doctors were gradually relying less on intangible "expert judgment" and trying instead to make treatment decisions based on data. The movement toward data-driven medicine, of course, needed to rest on good data and careful analysis. Often it didn't.

In one classic study, Ioannidis looked at papers from major medical journals that other researchers had cited at least a thousand times—a mark that they were having a major impact on the field. Of the forty-nine studies that met this criterion, seven had been flatly contradicted by further studies. Those included some famous mistakes in biomedicine, such as the claim that estrogen and progestin benefitted women who had had hysterectomies, when in fact the drug combination increased the risk of heart disease and breast cancer. Another ballyhooed study, which found that vitamin E reduced heart disease risk, turned out not to be true either. A few years later, Ioannidis followed up to see whether scientists were still citing that original, disproven study. The answer was yes—and frequently! Years after two of the largest and most expensive medical studies ever undertaken had debunked the claim that vitamin E reduces heart disease, half of all articles on the subject still cited the original study favorably. It left him shaking his head in disbelief. "How many trials of a billion dollars each can we do to refute a single claim out of the millions of claims that observational studies put forth?" he asked me. "We would need quintillions of dollars just to show what things are worthless before we start doing our real job," which is finding treatments that actually work. He suspects that scientists who had spent their careers studying vitamin E kept on defending the positive findings. "They were living in their own bubble, unperturbed by the evidence."

"This is one major reason why having lots of false results circulating in the literature is not a good idea. These results get entrenched. You cannot get rid of them," he told me. "There will also be lots of people who are unaware of it who will just hit upon the paper and will never know that this thing has been refuted."

Ioannidis's papers have now collectively garnered more than 100,000 citations. None is more widely known than his 2005 paper titled "Why Most Published Research Findings Are False." It didn't make the public splash that Glenn Begley's did, but it became a touchstone for many academic discussions about the foibles of biomedical research. The odd thing about the heavily cited paper is that it contains no hard data. There's no survey of researchers or random sampling of papers in the scientific literature. Instead, it's an essay in which Ioannidis makes a purely statistical argument. In essence, he concludes that simply by looking at how scientific research is designed and executed, he could tell that many papers were nothing more than false-positive results.

At first the paper caught the attention of statisticians and study designers. "Gradually more people started seeing these problems and being interested in these problems, and wanted to see what was happening in their field," Ioannidis said. The provocative title probably helped. "I think actually the title was a bit of a risk because if the paper didn't have substance then it would very easily backfire," Ioannidis said.

Goodman, for one, was not convinced by Ioannidis's statistical argument. He and a colleague, Sander Greenland, questioned its underlying assumptions. "We agree with the paper's conclusions and recommendations that many medical research findings are less definitive than readers suspect," Goodman and Greenland wrote, but "the claim that 'most research findings are false for most research designs and for most fields' must be considered as yet unproven."

In addition to the pushback Ioannidis got from Goodman, Jeff Leek at Johns Hopkins also published a critique, using actual data from top medical journals as input for his calculations. Leek's paper concluded that the failure rate in those publications was probably around 14 percent, which is not nearly as dire as the assertion that "most" findings are false. But that's not necessarily inconsistent with Ioannidis's essay, which noted that failure rates range from 15 to 99 percent or more, depending on the size and design of the study. Large clinical studies are the best, while smaller laboratory studies, based on small samples, are, statistically speaking, unlikely to be true most of the time. (Scientists do raise their eyebrows when you suggest that an entire class of small studies is wrong 99 percent of the time, as the Ioannidis calculation suggests.) Other scientists quoting his paper rarely mention that the success rate varies dramatically, depending on the study type. "I think that oversimplifying to just get an average is not very helpful," Ioannidis told me, though his paper's title of course encouraged readers to do just that. Whatever its flaws, there's no question that the paper helped stimulate the current conversation about irreproducibility in biomedical research and how to deal with it.

Scientists looking for ways to improve the reliability of laboratory research have much to learn from an earlier push to improve medical research involving human subjects. Over the past twenty years, there's been significant progress in that arena. The best clinical studies are now designed and carried out with great care (and at great expense). They involve many patients and often multiple research centers to make the findings more robust. And those improvements have helped push medical science forward, providing credible evidence for treating and managing disease, while gradually driving bad ideas out of practice. To cite one example, a careful study determined that hormone replacement therapy was deadly for many women taking the combination of estrogen and progestin, claiming many lives during the years doctors prescribed those drugs together. By one estimate, that corrective study triggered a change in medical practice that averted 126,000 breast cancer deaths and 76,000 heart disease fatalities between 2003 and 2012.

"I would be the last to say we've solved all the problems of clinical research," Goodman told me. "But at least we have a decent template of what needs to be done." The question is how to apply the lessons from clinical research to the world of laboratory science. There's no cookie-cutter solution. Each field of science has its own particular culture, so each will have to develop its own ways to improve rigor. For example, Goodman discovered that in psychology, a single experiment often becomes the basis for an entire career, and replication is actually discouraged. "To redo an experiment is taken as a personal attack on the integrity and on the theories of the person who did the original work," Goodman found. "I thought I couldn't be shocked, but this is truly shocking."

There's also a big difference between laboratory research, where scientists are trying to figure out the mechanism of disease, and drug testing in human subjects, where overarching questions are practical: Is a drug or procedure safe and effective? A clinical trial sets out to answer those comparatively straightforward yes or no questions. Lab research often explores the much more difficult question of why. As a result, clinical studies and basic biomedical research "are profoundly different cultures."

Importing solutions directly from clinical research and applying them in the laboratory is not likely to work. Goodman made an analogy to foreign aid projects, where Westerners parachute in and attempt to impose a solution, without fully understanding the local customs. "You can make it much worse if you don't respect the culture," he said. What's more, each area of biomedicine has its own insular subculture, in which ideas and methods—good and bad—circulate and reverberate. Norms change over time, but they don't transmit easily from one field to another.

Many of the needed improvements—randomizing animals in studies, keeping lab personnel unaware of which are test subjects versus comparison groups, not changing the end point once the experiment has started, starting with an adequate sample size—can be made unobtrusively. Often, lab scientists making mistakes in those areas "didn't know it was important," Goodman said. Now that the word is spreading, he expects this will start to change—leading eventually, he hopes, to a social transformation of biomedical science.

Goodman and Ioannidis are trying to accelerate that social transformation. To that end, in the fall of 2015, they stood before a crowd arranged around tables in an airy conference center on the Stanford campus. They had invited dozens of scientists from the United States and Europe who have studied the issue of rigor in biomedical research and asked them a provocative question: How can this nascent community chart a research agenda to study these systematic problems and identify and test potential solutions?

The discussion touched on four topics that generally arise when scientists think about how to fix the broken system: getting individual scientists to change their ways, getting journals to change their incentives, getting funding agencies to promote better practices, and, last but not least, getting universities to grapple with these issues. Of course these are interlocking challenges.

Take, for instance, the fact that universities rely far too heavily on the number of journal publications to judge scientists for promotion and tenure. Brian Nosek said that when he went up for promotion to full professor at the

University of Virginia, the administration told him to print out all his publications and deliver them in a stack. Being ten years into his career, he'd published about a hundred papers. "So my response was, what are you going to do? Weigh them?" He knew it was far too much effort for the review committee to read one hundred studies. "So the message that's being delivered to me was . . . volume matters." Nosek told me that delivering his three best papers would have provided a more meaningful view of his accomplishments to date. And if he'd known up front that the quality of his findings, rather than the sheer number of his papers, would be key to getting tenure, that "would totally change the incentives" for his course of research. He would have spent more time thinking about a few big and interesting questions to tackle rather than worrying about populating his list of publications.

Frank Miedema, dean of the medical school at the University Medical Center in Utrecht, Holland, looked at this issue from an institution's point of view. He complained that scientists at his medical center publish 3,500 papers a year, "and I don't know who reads them. Have you read one of our papers?" he asked the audience at Stanford. The answer was silence. The push for quantity is utterly misplaced, he argued. Scientists aren't asking questions with important answers; they're asking easily answerable questions. Nobody wants to risk spending four years on a risky research project with a big potential payoff, but that could also fall flat. That might be a powerful way to move medical science forward, but the risks to a scientist's career are enormous, given the current incentive structure.

Miedema said he's trying an experiment at his institution to break out of that academic trap. "Ask the patients, and they tell you what they want. That's what we do." His medical school judges the scientists there on the public impact of their work and cares less about curiosity-driven studies. "Most papers are never used, and rarely read," he said. "If we don't incentivize and reward people to do the right things, they will not do the right things, and they will keep on publishing this waste, and nothing will change," he said. "You guys will be here ten years from now, and John [Ioannidis] will have even less hair . . . and there will be no change in the system."

Robert Califf, then awaiting confirmation as Food and Drug Administration (FDA) commissioner, said that sensibility is starting to take hold in the United States as well. "Academia has to clean up its shop and get out of the ego business and get into the business of answering questions that matter to patients," he said. "But the beauty is the patients are gradually going to be taking control." If academics don't answer the questions the public cares about, "there's a very high chance you won't get funded because they're going to have a lot to say about it." Politicians already steer some biomedical research dollars through the Defense Department, which is heavily influenced by patient advocacy groups that participate in the peer review process. Those groups are trying to get more involved in setting the research agenda among scientists funded by the National Institutes of Health (NIH) as well. But they also need cooperation from scientists, who are used to dreaming up their own ideas and getting grants to follow their intellectual muses. Both kinds of science are essential; it's a question of balance.

Incentives to improve academic research can also come from pharmaceutical companies, which have whittled away their own research departments and depend increasingly on academia for new product leads. Glenn Begley's shot across the bow in 2012 laid out the problem in stark terms. Some universities are now pursuing research that blends pure scientific exploration and commercialization. Barbara Slusher is pioneering one of those efforts. She has been on both sides of the divide and runs an operation at Johns Hopkins University to take the best from each approach. Academics sometimes disdain industry scientists as unimaginative. But industry gets some important things right. And as they're preparing to move a drug through the FDA and toward market, they must follow FDA-approved good laboratory practice guidelines, which adds a layer of bureaucracy (largely careful documentation) to their work.

Slusher tries to use some of the tools of industry to validate ideas from academic labs—before they head toward drug development. She said she's trying to avoid more papers like Glenn Begley's, in which pharma complains about the poor quality of academic research. "It's not good. Not good. So our thought is let's keep it in-house. Let's keep it within the family" before the ideas go on to pharma. "If you talk about solutions, I think that's something we'll see a lot more of."

Even at Hopkins, one of the nation's top research institutions, her lab often struggles to reproduce results from the university's labs. They've explored dozens of promising ideas. "I'd say we've had better than 50 percent reproduction, and I think that's probably because we're working hand in hand with the faculty that made the initial discovery," she told me. "We've got to get rid of this irreproducibility issue. That's a problem. We've got to get better."

Glenn Begley has weighed in on this as well. He and two colleagues wrote that good laboratory practice, which works well in industry, should be adapted for academia. "The scientific community should come up with a similar system for research, which we term good institutional practice (GIP). If funding depended on a certified record of compliance with GIP, robust research would get due recognition." Michael Rosenblatt from Merck has suggested an even more aggressive remedy: drug companies should fund more research at universities, but, in exchange, universities should offer a money-back guarantee if the findings don't hold up. That would obviously make universities take a more active role in ensuring the reproducibility of research conducted within their walls.

Journal publishers could also play a role in easing the problems of reproducibility. One simple step would be to publish more studies that report "negative results," that is, that fail to replicate a previously reported positive finding. High-profile journals are reluctant to do that now because those follow-up studies get cited less frequently than new and exciting ones, potentially reducing a publication's impact factor and therefore its profits. Daniele Fanelli at METRICS has also suggested that journals set up a system of "self-retraction" so that scientists who find honest errors can flag them in the journal that published the original work. You'd think that would already happen, but in fact, because retractions are often assumed to be the result of questionable behavior, scientists are loath to admit honest errors. Colleagues wonder about the backstory, which can harm reputations and careers. Retractions "are often a source of dispute among authors and a legal headache for journal editors," Fanelli wrote. These self-retractions would be signed by all authors as a signal that they were the result of honest error. He went on to suggest that journals should consider a year of "scientific jubilee" during which papers could be self-retracted, no questions asked. "The literature would be purged, repentant scientists would be rewarded, and those who had sinned, blessed with a second chance, would avoid future temptation."

There's also plenty of room to improve the journals' own peer review process. Steve Goodman was startled to discover that *Science* magazine didn't have a formal board of statistics editors until 2015 (though it did use statisticians as reviewers before then). "This has been recognized as absolutely critical to the review of empirical science for decades.

And yet *Science* magazine just figured it out. How could that be?" Another problem: peer review is usually unpaid, so scientists may delegate the job to graduate students or spend less time than it might take to uncover problems with a paper.

Many scientists aren't waiting for journals to change their ways. Social media has created many avenues for scientists to carry on these conversations outside the traditional channels. Scientific firebrands like Michael Eisen at the University of California (UC), Berkeley, tweet out 140-character critiques of their colleagues' work. Paul Knoepfler at UC-Davis writes pointed blog posts about research that concerns him. Other scientists are posting comments on a site called PubPeer, which allows them to take anonymous potshots at research articles. And the NIH has gotten in on the act as well, creating Pubmed Commons, an on-the-record comment section connected to the main publications database.

A British organization, the Faculty of 1000, started a Preclinical Reproducibility and Robustness channel on its website, which accepts papers that critique or describe failed replications of previous studies. Articles are posted, and peer review comes in the form of comments. Scientists have also started posting to bioRxiv.org, a "preprint" site that doesn't require peer review upfront but counts on scientists who comment on those papers to serve in place of journal gatekeepers. These movements could eventually devalue the journal article as the ultimate currency of scientific research and move toward a more fluid world where the record evolves along with the science. Ahmed Alkhateeb, a postdoc at Harvard Medical School, wrote an opinion piece suggesting that scientists should publish more bite-sized bits of research, with greater focus on a small increment of new data and less focus on the analysis that attempts to weave that new information into a broader scientific narrative. He argued that this system would reduce the incentive for scientists to focus unduly on data that support a popular hypothesis. A more nimble publication system might also encourage scientists to publish confirmatory or negative results.

Marcia McNutt, who spoke at the Stanford meeting as editor in chief of *Science* (before becoming president of the National Academy of Sciences), worried that deemphasizing the role of journals as gatekeepers would make it even more difficult for young scientists and students to know what to trust in the literature. At the same time, she acknowledged the limits of scientific publication by recounting a story about John Maddox, longtime editor in chief of *Nature*. Someone once asked him how much of what *Nature* published was wrong, "and he famously answered, 'All of it,'" McNutt said. "What he meant by that is, viewed through the lens of time, just about everything that we write down we'll look back at and say, 'That isn't quite right. That doesn't really look like how we would express things today.' So most papers don't stand up to the test of time." McNutt herself said that of the papers deep in *Science*'s archives, "I probably wouldn't publish them again today, even if I didn't care about how up-to-date they were." Her point wasn't that journals are useless, of course, but that scientific findings are provisional and should be treated as such.

Many of these suggestions have financial implications, whether for journals, which could lose stature if they publish less flashy papers, or for universities, which might have to offer money-back guarantees to funders from industry. But Brian Nosek argued that money isn't the only way to change human behavior. "The solutions don't require a huge shift in budget," he said at the Stanford meeting. "They require a small shift in budget." And sometimes small incentives can have an outsized impact. One idea he has pursued involves awarding "badges" to scientists who do the right thing. Like a gold star on an elementary school assignment, these visible tokens mark published papers whose authors have agreed to share their data. "Badges are stupid. But they work," he said.

The Center for Open Science ran an analysis after the journal *Psychological Science* started publishing openness badges in 2014. Though many scientists who published in the journal didn't seek this goody-goody mark of approval, their behavior changed nonetheless. A year after the journal started posting badges, the percentage of papers with open data rose from 3 to 38 percent, Nosek and his colleagues found.

Of course, reproducibility would improve if scientists took simple technical steps, such as validating cell lines, running proper controls with their antibody experiments, choosing adequate sample sizes for mouse studies, deciding in advance what hypothesis they were testing, and so on.

Scientists like Nosek hope to make those practices more common simply by raising awareness. One vehicle for doing that is publishing guidelines and checklists for scientists to follow. The ARRIVE guidelines, for example, provide a template for scientists who publish results from animal experiments. Nosek convened a committee that developed the Transparency and Openness Promotion (TOP) guidelines. A survey of animal-research guidelines in 2013 identified twenty-six distinct sets, including fifty-five specific recommendations (such as randomization and adequate sample size). Major journals have adopted publication guidelines, negotiated at an NIH-sponsored meeting, that they ask authors to follow. *Nature*, for example, requires scientists to complete a checklist stating whether they have authenticated their cell lines—but the journal may still publish a "hot" paper even if scientists haven't fully complied. And even the most widely accepted guidelines are frequently ignored.

That leads to an obvious conclusion: awareness isn't enough. The social context of science needs to change in order to create the incentives that will lead scientists to raise their standards. Yet most of the scientists who are trying to fix the problems of reproducibility are biologists or physicians, not social scientists equipped to think about remodeling the culture of science. Social scientists who pay attention to the study of scientific research tend to work on esoteric topics rather than the more nuts-and-bolts issues involved in understanding and changing ongoing behaviors. But a few pioneers, like Nosek and Brian Martinson, have pushed into this territory. Jonathan Kimmelman, who focuses on biomedical ethics at McGill University in Montreal, is another. At the Stanford meeting, he challenged researchers to think more deeply about modifying scientists' behavior.

Kimmelman argued provocatively that since science can never free itself of missteps and irreproducible results, it would be helpful for scientists, when they report a result, to state how much confidence they have in their findings. If it's a wild idea, declare that you don't have a whole lot of confidence in the result, and scientists following up on it can proceed at their own risk. If you're very confident of your result, say so. And if you have a good track record, that will instill confidence in your findings. Of course this system will only work if these subjective judgments are better than the flip of a coin. Kimmelman has been running experiments to measure how well scientists can, in fact, make these predictions.

Scientists make judgments all the time, not only about their own work but about the papers they read. Kimmelman hopes that these judgments can be quantified and reported as a matter of course. With this strategy, Kimmelman is trying to take advantage of human abilities that are not conveyed in the dry analysis of journal articles. It's "getting to what's going on in the heads of people," he told me. "That's not only one of the missing pieces in the puzzle here, but I think it's a really, really critical issue."

During our conversation he floated an idea that seemed almost heretical: maybe a certain amount of error is necessary, because it gives scientists something to argue over. The stock market wouldn't work if everyone was in complete agreement about the value of a given share of stock. Nobody would buy or sell anything. "As with every economy, you may need a lot of riffraff" in science, Kimmelman suggested. This idea comes from looking at biomedical research as an interwoven system. Right now, he argued, "there's too much emphasis on the individual. What matters is where a community is going, not a particular lab," he said. "I don't mean to come off as an apologist at all. I'm not," he said. "In fact in most settings I guess I would be considered a critic. But I think there are a lot of aspects about reproducibility that we don't really understand well conceptually. I just think there's still further work to be done to clarify those kinds of things."

One final idea sounds downright counterintuitive: to speed the development of medicine, biomedical science should actually slow down. This means taking on fewer projects and doing them more carefully. It means improving the quality of the scientific literature by publishing fewer, more careful papers. In 1963, physicist-historian Derek de Solla Price warned that the scientific literature was growing exponentially and would eventually become unmanageable unless something was done to change the incentives in science. Daniel Sarewitz at Arizona State University wrote that Price's premonition is coming true. "Today, the interrelated problems of scientific quantity and quality are a frightening manifestation of what he foresaw. It seems extraordinarily unlikely that these problems will be resolved through the home remedies of better statistics and lab practice, as important as they may be." Given the current reality, Sarewitz told me, scientists and the public would be better off if we actually expected *less* from science.

We should not assume that every paper in the literature falls neatly into the "good" basket or the "bad" basket. Much of it is provisional, and its true worth may take decades to appreciate. Some of today's medical advances stem from discoveries made decades ago, and presumably some of today's discoveries will prove valuable only many years from now. If we curb our enthusiasm a bit, scientists will be less likely to run headlong after dubious ideas like transdifferentiation, and the public will be less likely to embrace the latest dietary fad. Of course this is a discouraging point of view for patients and advocates looking for rapid progress in the search for treatments and cures. But it's important to distinguish between speed and haste.

A focus on quality over quantity would give Arturo Casadevall's students at Johns Hopkins a chance to think instead of simply running another experiment. It suggests a path (not without pain, alas) out of the structural morass that has made the biomedical research system financially unsustainable, with too many labs competing for the available funding. And in the end it speaks to a value that many scientists hold deeply: being right should matter most of all.

ACKNOWLEDGMENTS

Writing a book may seem like a solitary pursuit, but in my case it was deeply embedded in a social structure involving people who are also passionate about the topic at hand, or at least endowed with patience and generosity beyond the call. My agent, James Levine, helped me navigate the mysterious publishing world and gave me leave to tend to my writing without fretting about the process of turning my ideas into ink on pulp. Jim told me that my editor at Basic Books, T. J. Kelleher, was one of the best in the business when it comes to editing science books. I'm prepared to believe him. I thank T. J. for his cogent comments as the project took shape and his sharp edits. My editor at NPR, Joe Neel, also made many perceptive comments on a draft. I'm also grateful to Anne Gudenkauf, who runs the NPR science desk, for affording me the time to work on this project.

Most people quoted directly in this book talked to me in 2015 and 2016. Many gave me at least an hour of their valuable time, some considerably more. I'm grateful to all of them. I won't list their names here since you have already met them in the course of the book, but I would particularly like to thank Richard Neve and John Porter for their ongoing conversations with me. Olaf Andersen, Mark Winey, and members of his lab graciously put up with a barrage of questions while providing me hospitality. Thanks also to Roger Peng at Johns Hopkins, who helped me sort through statistical matters. I'm especially grateful to Tom Murphy, Sally Curtin, and Stan Artman, who shared their personal stories.

All quotes from the *Nature* journals are reprinted by permission from Macmillan Publishers Ltd. I use the copyrighted quotes from Stuart Firestein's *Failure: Why Science Is So Successful* with the permission of Oxford University Press. I'm grateful that the Public Library of Science journals and *eLife* make their material freely available through the Creative Commons Attribution license.

Last but not least, I am deeply grateful to Dan Sarewitz at the Consortium for Science, Policy & Outcomes of Arizona State University (ASU). He provided me a place to work in ASU's Washington, DC, offices, as well as an appointment as a visiting scholar, which included financial support. (Interviews in Tempe with Anna Barker, Carolyn Compton, and Josh LaBaer took place on ASU-supported trips.) Most importantly, I value Dan's many conversations with me as the book took shape and his insightful comments throughout.

RICHARD HARRIS,
September 2016

NOTES

Chapter 1: Begley's Bombshell

9 *The German drug maker Bayer:* Florian Prinz, Thomas Schlange, and Khusru Asadullah, "Believe It or Not: How Much Can We Rely on Published Data on Potential Drug Targets?," *Nature Reviews Drug Discovery* 10, no. 9 (2011): 712, doi:10.1038/nrd3439-c1.

10 *When the journal* Nature *published:* C. Glenn Begley and Lee M. Ellis, "Drug Development: Raise Standards for Preclinical Cancer Research," *Nature* 483, no. 7391 (2012): 531–533, doi:10.1038/483531a.

11 *In 2005, John Ioannidis published:* John P. A. Ioannidis, "Why Most Published Research Findings Are False," *PLOS Medicine* 2, no. 8 (2005), doi:10.1371/journal.pmed.0020124.

14 *He went on to calculate:* Leonard P. Freedman, Iain M. Cockburn, and Timothy S. Simcoe, "The Economics of

Reproducibility in Preclinical Research," *PLOS Biology* 13, no. 6 (2015): e1002165, doi:10.1371/journal.pbio.1002165.

15 *Two-thirds of the senior investigators:* Aaron Mobley et al., "A Survey on Data Reproducibility in Cancer Research Provides Insights into Our Limited Ability to Translate Findings from the Laboratory to the Clinic," *PLOS ONE* 8, no. 5 (2013): 3–6, doi:10.1371/journal.pone.0063221.

16 *The American Society for Cell Biology (ASCB):* "ASCB Member Survey on Reproducibility," ACSB, 2015, http://www.ascb.org/wp-content/uploads/2015/11/final-survey-results-without-Q11.pdf.

16 *From the director's office in Building 1:* Francis S. Collins and Lawrence A. Tabak, "Policy: NIH Plans to Enhance Reproducibility," *Nature* 505, no. 7485 (2014): 612–613, doi:10.1038/505612a.

18 *In 2012, Jack Scannell and his colleagues:* Jack W. Scannell et al., "Diagnosing the Decline in Pharmaceutical R&D Efficiency," *Nature Reviews Drug Discovery* 11, no. 3 (2012): 191–200, doi:10.1038/nrd3681.

23 *In 1999 and 2000, several scientists made a startling claim:* Timothy R. Brazelton et al., "From Marrow to Brain: Expression of Neuronal Phenotypes in Adult Mice," *Science* 290, no. 5497 (December 1, 2000): 1775–1779, http://science.sciencemag.org/content/290/5497/1775.abstract; E. Gussoni et al., "Dystrophin Expression in the Mdx Mouse Restored by Stem Cell Transplantation," *Nature* 401, no. 6751 (1999): 390–394, doi:10.1038/43919.

24 *In 2002, Wagers concluded with typical scientific:* A. J. Wagers et al., "Little Evidence for Developmental Plasticity of Adult Hematopoietic Stem Cells," *Science* 297, no. 5590 (2002): 2256–2259, doi:10.1126/science.1074807.

24 *"Most of these studies turned out":* Sean J. Morrison, "Time to Do Something About Reproducibility," *eLife* 3 (2014): 1–4, doi:10.7554/eLife.03981.

27 *His solution was to write a follow-up:* C. Glenn Begley, "Six Red Flags for Suspect Work," *Nature* 497 (2013): 433–434, doi:10.1038/497433a.

Chapter 2: It's Hard Even on the Good Days

29 *In the words of the brilliant physicist:* Richard P. Feynman, "Cargo Cult Science," Caltech, 1974, http://calteches.library.caltech.edu/51/2/CargoCult.htm.

30 *A natural scientist realized he could test:* David Wootton, *The Invention of Science: A New History of the Scientific Revolution* (New York: HarperCollins, 2015).

32 *"We might think of an experiment":* Martin A. Schwartz, "The Importance of Indifference in Scientific Research," *Journal of Cell Science* 128, no. 15 (2015): 2745–2746, doi:10.1242/jcs.174946. Quoted with author's permission.

35 *Her former mentor Elizabeth Blackburn:* Tonya L. Jacobs et al., "Intensive Meditation Training, Immune Cell Telomerase Activity, and Psychological Mediators," *Psychoneuroendocrinology* 36, no. 5 (2011): 664–681, doi:10.1016/j.psyneuen.2010.09.010.

35 *Steven Artandi and colleagues at Stanford University:* Jae-Il Park et al., "Telomerase Modulates Wnt Signalling by Association with Target Gene Chromatin," *Nature* 460, no. 7251 (2009): 66–72, doi:10.1038/nature08137.

36 *Based on that observation:* Margaret A. Strong et al., "Phenotypes in mTERT^{+}/$^{-}$ and mTERT^{-}/$^{-}$ Mice Are Due to Short Telomeres, not Telomere-Independent Functions of Telomerase Reverse Transcriptase," *Molecular and Cellular Biology* 31, no. 12 (2011): 2369–2379, doi:10.1128/MCB.05312-11.

36 *For example, his team discovered:* Linghe Xi and Thomas R. Cech, "Inventory of Telomerase Components in Human Cells Reveals Multiple Subpopulations of hTR and hTERT," *Nucleic Acids Research* 42, no. 13 (2014): 8565–8577, doi:10.1093/nar/gku560.

39 *"This has been characterized as":* Stuart Firestein, *Failure: Why Science Is So Successful* (Oxford: Oxford University Press, 2016), 159.

41 *Surveying papers from biomedical science:* David Chavalarias and John P. A. Ioannidis, "Science Mapping Analysis Characterizes 235 Biases in Biomedical Research," *Journal of Clinical Epidemiology* 63, no. 11 (2010): 1205–1215, doi:10.1016/j.jclinepi.2009.12.011.

41 *Only many years later did they appreciate:* J. A. Layton and F. S. Collins, "Policy: NIH to Balance Sex in Cell and Animal Studies," *Nature* 509, no. 7500 (2014): 282–283, doi:10.1038/509282a.

43 *The boys with the cash incentive:* L. N. Alfano et al., "T.P.1," *Neuromuscular Disorders* 24, no. 9–10 (2014): 860, doi:10.1016/j.nmd.2014.06.224.

47 *They published their tale:* William C. Hines et al., "Sorting Out the FACS: A Devil in the Details," *Cell Reports* 6, no. 5 (2014): 779–781, doi:10.1016/j.celrep.2014.02.021.

48 *"They just came to opposite conclusions":* Daniel H. Madsen and Thomas H. Bugge, "The Source of Matrix-Degrading Enzymes in Human Cancer: Problems of Research Reproducibility and Possible Solutions," *Journal of Cell Biology* 209, no. 2 (2015): 195–198, doi:10.1083/jcb.201501034.

49 *In the 1990s, pharmaceutical companies had spent:* Lisa M. Coussens, Barbara Fingleton, and Lynn M. Matrisian, "Matrix Metalloproteinase Inhibitors and Cancer: Trials and Tribulations," *Science* 295, no. 5564 (2002): 2387–2392, doi:10.1126/science.1067100.

Chapter 3: A Bucket of Cold Water

56 *His 2008 study shocked:* Sean Scott et al., "Design, Power, and Interpretation of Studies in the Standard Murine Model of ALS," *Amyotrophic Lateral Sclerosis: Official Publication of the World Federation of Neurology Research Group on Motor Neuron Diseases* 9, no. 1 (2008): 4–15, doi:10.1080/17482960701856300.

57 *The results: fail, fail, fail:* Steve Perrin, "Make Mouse Studies Work," *Nature* 507 (2014): 423, doi:10.1038/507423a.

59 *She started writing and talking:* Story C. Landis et al., "A Call for Transparent Reporting to Optimize the Predictive Value of Preclinical Research," *Nature* 490, no. 7419 (2012): 187–191, doi:10.1038/nature11556.

59 *"This is a great concern":* "Departments of Labor, Health and Human Services, and Education, and Related Agencies Appropriations for Fiscal Year 2013," US Government Publishing Office, https://www.gpo.gov/fdsys/pkg/CHRG-112shrg29104500/html/CHRG-112shrg29104500.htm.

60 *The two acknowledged the issue:* Francis S. Collins and Lawrence A. Tabak, "Policy: NIH Plans to Enhance Reproducibility," *Nature* 505, no. 7485 (2014): 612–613, doi:10.1038/505612a.

60 *And scientists must show:* "Rigor and Reproducibility," NIH, http://grants.nih.gov/reproducibility/index.htm.

68 *In fact, they've developed:* Jacqueline G. O'Rourke et al., "C9orf72 BAC Transgenic Mice Display Typical Pathologic Features of ALS/FTD," *Neuron* 88, no. 5 (2015): 892–901, doi:10.1016/j.neuron.2015.10.027; Owen M. Peters et al., "Human C9ORF72 Hexanucleotide Expansion Reproduces RNA Foci and Dipeptide Repeat Proteins but not Neurodegeneration in BAC Transgenic Mice," *Neuron* 88, no. 5 (2015): 902–909, doi:10.1016/j.neuron.2015.11.018.

Chapter 4: Misled by Mice

72 *Five people eventually died:* Institute of Medicine (US) Committee to Review the Fialuridine (FIAU/FIAC) Clinical

Trials, *Review of the Fialuridine (FIAU) Clinical Trials*, ed. F. J. Manning and M. Swartz (Washington, DC: National Academies Press, 1995).

72 *He asked a colleague:* Akira Endo, "A Historical Perspective on the Discovery of Statins," *Proceedings of the Japan Academy. Series B, Physical and Biological Sciences* 86, no. 5 (2010): 484–493, doi:10.2183/pjab.86.484.

72 *For instance, certain drug-toxicity tests:* Thomas Hartung, "Food for Thought; Look Back in Anger—What Clinical Studies Tell Us About Preclinical Work," *Altex* 30, no. 3 (2013): 275–291, doi:10.1016/j.biotechadv.2011.08.021.

74 *Neurologists started calling this long string:* Ulrich Dirnagl and Malcolm R. Macleod, "Stroke Research at a Road Block: The Streets from Adversity Should Be Paved with Meta-analysis and Good Laboratory Practice," *British Journal of Pharmacology* 157, no. 7 (2009): 1154–1156, doi:10.1111/j.1476-5381.2009.00211.

75 *It was a dramatic failure:* A. Shuaib et al., "NXY-059 for the Treatment of Acute Ischemic Stroke," *New England Journal of Medicine* 357, no. 6 (2007): 562–571, http://www.nejm.org/doi/full/10.1056/NEJMoa070240#t=article.

75 *Macleod dissected that study:* Dirnagl and Macleod, *British Journal of Pharmacology* (2009).

75 *"It is sobering":* Malcolm R. Macleod et al., "Risk of Bias in Reports of In Vivo Research: A Focus for Improvement," *PLOS Biology* 13, no. 10 (2015): 1–12, doi:10.1371/journal.pbio.1002273.

78 *The biology of inflammation:* Junhee Seok et al., "Genomic Responses in Mouse Models Poorly Mimic Human Inflammatory Diseases," *Proceedings of the National Academy of Sciences of the United States of America* 110, no. 9 (2013): 3507–3512, doi:10.1073/pnas.1222878110.

78 *"You can suppress things":* Keizo Takao and Tsuyoshi Miyakawa, "Genomic Responses in Mouse Models Greatly Mimic Human Inflammatory Diseases," *Proceedings of the National Academy of Sciences of the United States of America* 112, no. 4 (January 27, 2015): 1167–1172, doi:10.1073/pnas.1401965111.

78 *David Masopust at the University of Minnesota:* Lalit K. Beura et al., "Normalizing the Environment Recapitulates Adult Human Immune Traits in Laboratory Mice," *Nature* 532, no. 7600 (April 28, 2016): 512–516, http://dx.doi.org/10.1038/nature17655.

80 *Even so, these "identical" tests:* S. H. Richter et al., "Effect of Population Heterogenization on the Reproducibility of Mouse Behavior: A Multi-laboratory Study," *PLOS ONE* 6, no. 1 (2011), doi:10.1371/journal.pone.0016461.

80 *Even a man's sweaty T-shirt:* Robert E. Sorge et al., "Olfactory Exposure to Males, Including Men, Causes Stress and Related Analgesia in Rodents," *Nature Methods* 11, no. 6 (2014): 629–632, doi:10.1038/nmeth.2935.

81 *Garner goes even further in his thinking:* Joseph P. Garner, "The Significance of Meaning: Why Do over 90% of Behavioral Neuroscience Results Fail to Translate to Humans, and

What Can We Do to Fix It?," *ILAR Journal* 55, no. 3 (2014): 438–456, doi:10.1093/ilar/ilu047.

84 *And Hartung has private money:* Shraddha Chakradhar, "New Company Aims to Broaden Researchers' Access to Organoids," *Nature Medicine* 22, no. 4 (April 2016): 338, http://dx.doi.org/10.1038/nm0416-338.

86 *She showed me a video:* Kambez H. Benam et al., "Small Airway-on-a-Chip Enables Analysis of Human Lung Inflammation and Drug Responses in Vitro," *Nature Methods* 13, no. 2 (February 2016): 151–157, http://dx.doi.org/10.1038/nmeth.3697.

91 *Second, it's important to remember:* Jack W. Scannell and Jim Bosley, "When Quality Beats Quantity: Decision Theory, Drug Discovery, and the Reproducibility Crisis," *PLOS ONE* 11, no. 2 (2016), doi:10.1371/journal.pone.0147215.

Chapter 5: Trusting the Untrustworthy

93 *The team said it had isolated:* N. N. Desai et al., "Novel Human Endometrial Cell Line Promotes Blastocyst Development," *Fertility and Sterility* 61, no. 4 (1994): 760–766, http://www.ncbi.nlm.nih.gov/pubmed/7512055.

93 *As she described in a 2008 paper:* Nina Desai et al., "Live Births in Poor Prognosis IVF Patients Using a Novel Noncontact Human Endometrial Co-culture System," *Reproductive Biomedicine Online* 16, no. 6 (2008): 869–874, doi:10.1016/S1472-6483(10)60154-X.

96 *Even so, more than 7,000 published studies:* Jill Neimark, "Line of Attack," *Science* 347, no. 6225 (2015): 938–940, doi:10.1126/science.347.6225.938.

96 *A 2007 study estimated:* Peyton Hughes et al., "The Costs of Using Unauthenticated, Over-Passaged Cell Lines: How Much More Data Do We Need?," *BioTechniques* 43, no. 5 (2007): 575–586, doi:10.2144/000112598.

96 *"Have the Marx Brothers taken over":* Roland M. Nardone, "Curbing Rampant Cross-Contamination and Misidentification of Cell Lines," *BioTechniques* 45, no. 3 (2008): 221–227, doi:10.2144/000112925.

98 *Their list of contaminated cell lines:* Amanda Capes-Davis et al., "Check Your Cultures! A List of Cross-Contaminated or Misidentified Cell Lines," *International Journal of Cancer* 127, no. 1 (2010): 1–8, doi:10.1002/ijc.25242.

99 *The cells from this young woman:* R. Cailleau, M. Olive, and Q. V. Cruciger, "Long-Term Human Breast Carcinoma Cell Lines of Metastatic Origin: Preliminary Characterization," *In Vitro* 14, no. 11 (1978): 911–915, http://www.ncbi.nlm.nih.gov/pubmed/730202.

100 *In March 2000, Ross and his colleagues:* D. T. Ross et al., "Systematic Variation in Gene Expression Patterns in Human Cancer Cell Lines," *Nature Genetics* 24, no. 3 (2000): 227–235, doi:10.1038/73432.

101 *The NCI put up a note:* "MDA-MB-435, and Its Derivation MDA-N, Are Melanoma Cell Lines, Not Breast Cancer Cell Lines," Developmental Therapeutics Program, last updated

May 8, 2015, https://dtp.cancer.gov/discovery_development/nci-60/mda-mb-435.htm.

106 *For example, a study in Belgium:* Caroline Piette et al., "The Dexamethasone-Induced Inhibition of Proliferation, Migration, and Invasion in Glioma Cell Lines Is Antagonized by Macrophage Migration Inhibitory Factor (MIF) and Can Be Enhanced by Specific MIF Inhibitors," *Journal of Biological Chemistry* 284, no. 47 (2009): 32483–32492, doi:10.1074/jbc.M109.014589.

107 *The result? Scientists continue to publish:* Anja Torsvik et al., "U-251 Revisited: Genetic Drift and Phenotypic Consequences of Long-Term Cultures of Glioblastoma Cells," *Cancer Medicine* 3, no. 4 (2014): 812–824, doi:10.1002/cam4.219.

107 *Biologists in Uppsala, Sweden, isolated it:* Elie Dolgin, "Venerable Brain-Cancer Cell Line Faces Identity Crisis," *Nature*, August 31, 2016, doi:10.1038/nature.2016.20515.

107 *In 2016, scientists from Sweden decided:* Marie Allen et al., "Origin of the U87MG Glioma Cell Line: Good News and Bad News," *Science Translational Medicine* 8, no. 354 (August 31, 2016): 354re3–354re3, http://stm.sciencemag.org/content/8/354/354re3.abstract.

109 *"It basically didn't pan out":* Jean-Pierre Gillet, Sudhir Varma, and Michael M. Gottesman, "The Clinical Relevance of Cancer Cell Lines," *Journal of the National Cancer Institute* 105, no. 7 (2013): 452–458, doi:10.1093/jnci/djt007.

115 *After many years of increasingly excited effort:* Monya Baker, "Blame It on the Antibodies," *Nature* 521 (2015): 274–275, doi:10.1038/521274a.

116 *"Irisin," discovered in 2012:* Pontus Boström et al., "A PGC1 -α-Dependent Myokine That Drives Brown-Fat-Like Development of White Fat and Thermogenesis," *Nature* 481, no. 7382 (2012): 463–468, doi:10.1038/nature10777.

117 *He published a paper:* Elke Albrecht et al., "Irisin—a Myth Rather Than an Exercise-Inducible Myokine," *Scientific Reports* 5 (2015): 8889, doi:10.1038/srep08889.

117 *They published a follow-up paper:* Mark P. Jedrychowski et al., "Detection and Quantitation of Circulating Human Irisin by Tandem Mass Spectrometry," *Cell Metabolism* 22, no. 4 (2015): 734–740, doi:10.1016/j.cmet.2015.08.001.

118 *But David Rimm decided to call attention:* Jennifer Bordeaux et al., "Antibody Validation," *BioTechniques* 48, no. 3 (March 2010): 197–209, doi:10.2144/000113382.

120 *"As a result several thousand antibodies":* "1st International Antibody Validation Forum 2014: John Mountzouris," posted to YouTube by St John's Laboratory Ltd., October 30, 2014, https://www.youtube.com/watch?v=tnPUujPw2yY.

122 *Assuming scientists do step up:* Leonard P. Freedman, Iain M. Cockburn, and Timothy S. Simcoe, "The Economics of Reproducibility in Preclinical Research," *PLOS Biology* 13, no. 6 (2015): e1002165, doi:10.1371/journal.pbio.1002165.

Chapter 6: Jumping to Conclusions

123 *But Congress did just that:* "H.Con.Res.385—Expressing the Sense of the Congress That the Secretary of Health and Human Services Should Conduct or Support Research on Certain Tests to Screen for Ovarian Cancer, and Federal Health Care Programs and Group and Individual Health Plans Should Cover the Tests If Demonstrated to Be Effective, and for Other Purposes," Congress.gov, https://www.congress.gov/bill/107th-congress/house-concurrent-resolution/385.

123 *News of this putative new test:* Andrew Pollack, "New Cancer Test Stirs Hope and Concern," *New York Times*, February 3, 2004, http://www.nytimes.com/2004/02/03/science/new-cancer-test-stirs-hope-and-concern.html.

124 *Other scientists started raising doubts as well:* Mark Elwood, "Proteomic Patterns in Serum and Identification of Ovarian Cancer," *Lancet* 360, no. 9327 (July 13, 2002): 170, doi:http://dx.doi.org/10.1016/S0140-6736(02)09389-3.

126 *Analytical errors alone account for almost:* Leonard P. Freedman, Iain M. Cockburn, and Timothy S. Simcoe, "The Economics of Reproducibility in Preclinical Research," *PLOS Biology* 13, no. 6 (2015): e1002165, doi:10.1371/journal.pbio.1002165.

128 *"Common Genetic Variants Account for Differences":* Richard Spielman et al., "Common Genetic Variants Account for Differences in Gene Expression Among Ethnic Groups," *Nature Genetics* 39, no. 2 (2007): 226–231, doi:citeulike-article-id:1043226.

129 *The researchers wrote up a short analysis:* Joshua M. Akey et al., "On the Design and Analysis of Gene Expression Studies in Human Populations," *Nature Genetics* 39, no. 7 (July 2007): 807–808, http://dx.doi.org/10.1038/ng0707-807.

129 *These case studies became central examples:* Jeffrey T. Leek et al., "Tackling the Widespread and Critical Impact of Batch Effects in High-Throughput Data," *Nature Reviews Genetics* 11, no. 10 (2010): 733–739, doi:10.1038/nrg2825.

132 *He says only 1.2 percent:* John P. A. Ioannidis, Robert Tarone, and Joseph K. McLaughlin, "The False-Positive to False-Negative Ratio in Epidemiologic Studies," *Epidemiology* 22, no. 4 (2011): 450–456, doi:10.1097/EDE.0b013e31821b506e.

133 *He's found that 70 percent:* John P. A. Ioannidis et al., "The Geometric Increase in Meta-analyses from China in the Genomic Era," *PLOS ONE* 8, no. 6 (June 12, 2013): e65602, http://dx.doi.org/10.1371%2Fjournal.pone.0065602.

135 *As the story goes:* David Salsburg, *The Lady Tasting Tea: How Statistics Revolutionized Science in the Twentieth Century* (New York: W. H. Freeman, 2001).

136 *In the winter of 2015:* Michelle Schwalbe, "Statistical Challenges in Assessing and Fostering the Reproducibility of Scientific Results: Summary of a Workshop," National Academies Press," 2016, doi:10.17226/21915.

138 *In 2016, the American Statistical Association:* Ronald L. Wasserstein and Nicole A. Lazar, "The ASA's Statement on P-Values: Context, Process, and Purpose," *American*

Statistician 70, no. 2 (2016), doi:10.1080/00031305.2016.1154108.

139 *In a widely read 2011 paper:* Joseph P. Simmons, Leif D. Nelson, and Uri Simonsohn, "False-Positive Psychology: Undisclosed Flexibility in Data Collection and Analysis Allows Presenting Anything as Significant," *Psychological Science* 22, no. 11 (2011): 1359–1366, doi:10.1177/0956797611417632.

140 *In science, the equivalent practice:* Norbert L. Kerr, "HARKing: Hypothesizing After the Results Are Known," *Personality and Social Psychology Review* 2, no. 3 (1998): 196–217, doi:10.1207/s15327957pspr0203_4.

143 *The judge in the case agreed:* "AIDS Healthcare Foundation, Plaintiff, vs. United States Food and Drug Administration, et al.," AIDS Healthcare Foundation, http://www.aidshealth.org/wp-content/uploads/2013/06/Doc-60-Order-Denying-FDAs-MSJ.pdf.

Chapter 7: Show Your Work

146 *The results made news around the world:* Open Science Collaboration, "Estimating the Reproducibility of Psychological Science," *Science* 349, no. 6251 (2015): aac4716–aac4716, doi:10.1126/science.aac4716.

148 *He hectors journals to publish:* See his website at http://www.alltrials.net.

148 *Only 8 percent of the studies:* Robert M. Kaplan and Veronica L. Irvin, "Likelihood of Null Effects of Large NHLBI Clinical Trials Has Increased over Time," *PLOS ONE* 10, no. 8 (2015), doi:10.1371/journal.pone.0132382.

151 *The paper has been cited:* E. S. Lander et al., "Initial Sequencing and Analysis of the Human Genome," *Nature* 409, no. 6822 (2001): 860–921, doi:10.1038/35057062.

152 *Three months later, Salzberg and his colleagues:* S. L. Salzberg et al., "Microbial Genes in the Human Genome: Lateral Transfer or Gene Loss?," *Science* 292, no. 5523 (2001): 1903–1906, doi:10.1126/science.1061036.

153 *Salzberg was posting his data:* Bjorn Nystedt et al., "The Norway Spruce Genome Sequence and Conifer Genome Evolution," *Nature* 497, no. 7451 (May 30, 2013): 579–584, http://dx.doi.org/10.1038.

156 *Brian Nosek paired up with:* Timothy M. Errington et al., "An Open Investigation of the Reproducibility of Cancer Biology Research," ed. Peter Rodgers, *eLife* 3 (2014): e04333, doi:10.7554/eLife.04333.

159 *He was senior author of one:* C. L. Chaffer et al., "Normal and Neoplastic Nonstem Cells Can Spontaneously Convert to a Stem-Like State," *Proceedings of the National Academy of Sciences of the United States of America* 108, no. 19 (2011): 7950–7955, doi:10.1073/pnas.1102454108.

Chapter 8: A Broken Culture

170 *"I rather hate the idea":* Francis Darwin, ed., *The Life and Letters of Charles Darwin* (London: D. Appleton and Co., 1898), 427. Accessed via Google Books.

171 *Hearing the hoofbeats of competition:* Kathleen Collins, Ryuji Kobayashi, and Carol W. Greider, "Purification of Tetrahymena Telomerase and Cloning of Genes Encoding the Two Protein Components of the Enzyme," *Cell* 81, no. 5 (1995): 677–686, doi:10.1016/0092-8674(95)90529-4.

171 *Soon thereafter, Lingner and his mentor:* J. Lingner and T. R. Cech, "Purification of Telomerase from *Euplotes Aediculatus:* Requirement of a Primer 3' Overhang," *Proceedings of the National Academy of Sciences of the United States of America* 93, no. 20 (1996): 10712–10717, http://www.pnas.org/content/93/20/10712.short. Stephan, now at Stanford, came to a similar conclusion.

171 *She wrote another paper:* Chantal Autexier, D. X. Mason, and C. W. Greider, "Tetrahymena Proteins p80 and p95 Are Not Core Telomerase Components," *Proceedings of the National Academy of Sciences of the United States of America* 98, no. 22 (2001): 12368–12373, doi:10.1073/pnas.221456398.

173 *A study by the National Institutes of Health:* "Biomedical Research Workforce Working Group Report," NIH, June 14, 2012, http://acd.od.nih.gov/Biomedical_research_wgreport.pdf, p. 81.

177 *On the December day in 2013:* Randy Schekman, "How Journals like Nature, Cell and Science Are Damaging Science," *Guardian*, December 9, 2013, http://www.theguardian.com/commentisfree/2013/dec/09/how-journals-nature-science-cell-damage-science.

179 *Researchers in Japan claimed to have:* Haruko Obokata et al., "Stimulus-Triggered Fate Conversion of Somatic Cells into Pluripotency," *Nature* 505, no. 7485 (January 30, 2014): 641–647, http://dx.doi.org/10.1038/nature12968.

179 *The paper was reportedly rejected:* Gretchen Vogel and Dennis Normile, "Exclusive: Nature Reviewers Not Persuaded by Initial STAP Stem Cell Papers," *Science*, September 11, 2014, http://www.sciencemag.org/news/2014/09/exclusive-nature-reviewers-not-persuaded-initial-stap-stem-cell-papers.

180 *Investigators said she falsified dozens:* "Case Summary: Forbes, Meredyth M.," Office of Research Integrity, https://ori.hhs.gov/content/case-summary-forbes-meredyth-m.

180 *Investigators found that he "duplicated images":* "Case Summary: Pastorino, John G.," Office of Research Integrity, https://ori.hhs.gov/content/case-summary-pastorino-john-g.

181 *Robert Weinberg at the Massachusetts Institute of Technology:* "Cancer Research Retraction Is Fifth for Robert Weinberg; Fourth for His Former Student," Retraction Watch, http://retractionwatch.com/2015/07/06/cancer-research-retraction-is-fifth-for-robert-weinberg-fourth-for-his-former-student.

182 *Arturo Casadevall at Johns Hopkins University and colleague:* Ferric C. Fang, R. Grant Steen, and Arturo Cadadevall,

"Misconduct Accounts for the Majority of Retracted Scientific Publications," *Proceedings of the National Academy of Sciences of the United States of America* 109, no. 42 (2012): 17028–17033, doi:10.1073/pnas.1220833110.

182 *They were flabbergasted to find:* David B. Allison et al., "Reproducibility: A Tragedy of Errors," *Nature* 530 (February 3, 2016): 27–29, http://www.nature.com/news/reproducibility-a-tragedy-of-errors-1.19264.

184 *The original authors were given a chance:* Richard S. Spielman and Vivian G. Cheung, "Reply to 'On the Design and Analysis of Gene Expression Studies in Human Populations,'" *Nature Genetics* 39, no. 7 (2007): 808–809, doi:10.1038/ng0707-808.

186 *He has documented some of this behavior:* Brian C. Martinson, Melissa S. Anderson, and Raymond de Vries, "Scientists Behaving Badly," *Nature* 435, no. 7043 (2005): 737–738, doi:10.1038/435737a.

186 *Daniele Fanelli, now at Stanford:* Daniele Fanelli, "How Many Scientists Fabricate and Falsify Research? A Systematic Review and Meta-analysis of Survey Data," *PLOS ONE* 4, no. 5 (2009), doi:10.1371/journal.pone.0005738.

187 *"But if you feel the principles":* Brian C. Martinson et al., "The Importance of Organizational Justice in Ensuring Research Integrity," *Journal of Empirical Research on Human Research Ethics: JERHRE* 5, no. 3 (2010): 67–83, doi:10.1525/jer.2010.5.3.67.

188 *Martinson pointed to a paper:* Paul E. Smaldino and Richard McElreath, "The Natural Selection of Bad Science," *Royal*

Society Open Science 3, no. 9 (September 21, 2016), http://rsos.royalsocietypublishing.org/content/3/9/160384.abstract.

190 *Labor economist Paula Stephan at Georgia State University:* Paula Stephan, "The Endless Frontier: Reaping What Bush Sowed?," in *The Changing Frontier: Rethinking Science and Innovation Policy*, ed. Adam Jaffe and Benjamin Jones (Chicago: University of Chicago Press, 2015).

190 *They found a dramatic increase:* Christiaan H. Vinkers, Joeri K. Tijdink, and Willem M. Otte, "Use of Positive and Negative Words in Scientific PubMed Abstracts Between 1974 and 2014: Retrospective Analysis," *BMJ* 351 (2015): h6467, doi: http://dx.doi.org/10.1136/bmj.h6467.

192 *Maxim Shatsky and Richard Hall:* Maxim Shatsky et al., "A Method for the Alignment of Heterogeneous Macromolecules from Electron Microscopy," *Journal of Structural Biology* 166, no. 1 (2009): 67–78, doi:10.1016/j.jsb.2008.12.008.

192 *"One must not underestimate":* R. Henderson, "Avoiding the Pitfalls of Single Particle Cryo-electron Microscopy: Einstein from Noise," *Proceedings of the National Academy of Sciences of the United States of America* 110, no. 45 (2013): 18037–18041, doi:10.1073/pnas.1314449110.

193 *But these young scientists seemed:* K. T. Dolan, J. F. Pierre, and E. J. Heckler, "Revitalizing Biomedical Research: Recommendations from the Future of Research Chicago Symposium [version 1; referees: awaiting peer review]," *F1000Research* 5 (2016):1548, doi: 10.12688/f1000research.9080.1.

194 *"Rescuing US Biomedical Research from Its Systemic Flaws":* Bruce Alberts et al., "Rescuing US Biomedical Research from Its Systemic Flaws," *Proceedings of the National Academy of Sciences of the United States of America* 111, no. 16 (2014): 5773–5777, doi:10.1073/pnas.1404402111.

194 *Attendees did agree, though, on one point:* Bruce Alberts et al., "Opinion: Addressing Systemic Problems in the Biomedical Research Enterprise: Fig. 1," *Proceedings of the National Academy of Sciences of the United States of America* 112, no. 7 (2015): 1912–1913, doi:10.1073/pnas.1500969112.

Chapter 9: The Challenge of Precision Medicine

199 *And that was enough to degrade:* D. G. Hicks and L. Schiffhauer, "Standardized Assessment of the HER2 Status in Breast Cancer by Immunohistochemistry," *Laboratory Medicine* 42, no. 8 (2011): 459–467, doi:10.1309/LMGZZ58CTS0DBGTW.

199 *Two leading professional societies:* M. Elizabeth H. Hammond et al., "American Society of Clinical Oncology/college of American Pathologists Guideline Recommendations for Immunohistochemical Testing of Estrogen and Progesterone Receptors in Breast Cancer," *Journal of Clinical Oncology* 28, no. 16 (2010): 2784–2795, doi:10.1200/JCO.2009.25.6529.

201 *And it turned out that the hospital:* Josep Villanueva et al., "Correcting Common Errors in Identifying Cancer-Specific Serum Peptide Signatures," *Journal of Proteome Research* 4, no. 4 (2005): 1060–1072, doi:10.1021/pr050034b.

202 *With that in mind:* Francis S. Collins and Anna D. Barker, "Mapping the Cancer Genome," *Scientific American* 296, no. 3 (March 2007): 50–57, doi:10.1038/scientificamerican0307-50.

204 *In 2012, the group published:* Mathew J. Garnett et al., "Systematic Identification of Genomic Markers of Drug Sensitivity in Cancer Cells," *Nature* 483, no. 7391 (2012): 570–575, doi:10.1038/nature11005.

205 *The Broad team published its first findings:* Jordi Barretina et al., "The Cancer Cell Line Encyclopedia Enables Predictive Modelling of Anticancer Drug Sensitivity," *Nature* 483, no. 7391 (2012): 603–607, doi:10.1038/nature11003.

205 *The following year, they published:* Benjamin Haibe-Kains et al., "Inconsistency in Large Pharmacogenomic Studies," *Nature* 504, no. 7480 (2013): 389–393, doi:10.1038/nature12831.

206 *Two years later the authors:* Nicolas Stransky et al., "Pharmacogenomic Agreement Between Two Cancer Cell Line Data Sets," *Nature* (2015): 84–87, doi:10.1038/nature15736.

206 *The conflict spiraled:* Zhaleh Safikhani et al., "Assessment of Pharmacogenomic Agreement," *F1000Research* 5 (2016): 825, doi:10.12688/f1000research.8705.1.

207 *But that third analysis also focused:* Peter M. Haverty et al., "Reproducible Pharmacogenomic Profiling of Cancer Cell Line Panels," *Nature* 533, no. 7603 (May 19, 2016): 333–337, http://dx.doi.org/10.1038/nature17987.

210 *Sorger has been arguing:* Marc Hafner et al., "Growth Rate Inhibition Metrics Correct for Confounders in Measuring

Sensitivity to Cancer Drugs," *Nature Methods* 13, no. 6 (2016): 521–527, doi:10.1038/nmeth.3853.

214 *Barker's glioblastoma study, called GBM Agile:* "GBM AGILE," National Biomarker Development Alliance, http://nbdabiomarkers.org/gbm-agile.

214 *Researchers studying breast cancer pioneered:* Malorye Allison, "Biomarker-Led Adaptive Trial Blazes a Trail in Breast Cancer," *Nature Biotechnology* 28, no. 5 (2010): 383–384, doi:10.1038/nbt0510-383.

Chapter 10: Inventing a Discipline

219 *In one classic study:* John P. A. Ioannidis, "Contradicted and Initially Stronger Effects in Highly Cited Clinical Research," *JAMA: The Journal of the American Medical Association* 294, no. 2 (2005): 218–228, doi:10.1001/jama.294.2.218.

219 *Years after two of the largest:* Athina Tatsioni, Nikolaos G. Bonitsis, and John P. A. Ioannidis, "Persistence of Contradicted Claims in the Literature," *JAMA: The Journal of the American Medical Association* 298, no. 21 (2007): 2517–2526, doi:10.1016/j.jemermed.2008.02.043.

220 *"Why Most Published Research Findings Are False":* John P. A. Ioannidis, "Why Most Published Research Findings Are False," *PLOS Medicine* 2, no. 8 (2005), doi:10.1371/journal.pmed.0020124.

221 *"We agree with the paper's":* Steven Goodman and Sander Greenland, "Assessing the Unreliability of the Medical

Literature: A Response to 'Why Most Published Research Findings Are False,'" *PLOS Medicine* 4, no. 4 (2007): 135, doi:10.1371/journal.pmed.0040168.

221 *In addition to the pushback:* Leah R. Jager and Jeffrey T. Leek, "An Estimate of the Science-Wise False Discovery Rate and Application to the Top Medical Literature," *Biostatistics* 15, no. 1 (2014): 1–12, doi:10.1093/biostatistics/kxt007.

222 *By one estimate, that corrective study triggered:* Robert M. Kaplan and Veronica L. Irvin, "Likelihood of Null Effects of Large NHLBI Clinical Trials Has Increased over Time," *PLOS ONE* 10, no. 8 (2015), doi:10.1371/journal.pone.0132382.

228 *He and two colleagues wrote that good:* C. Glenn Begley, Alastair M. Buchan, and Ulrich Dirnagl, "Robust Research: Institutions Must Do Their Part for ility," *Nature* 525, no. 7567 (2015): 25–27, doi:10.1038/525025a.

228 *"If funding depended on a certified record":* Michael Rosenblatt, "An Incentive-Based Approach for Improving Data Reproducibility," *Science Translational Medicine* 8, no. 336 (April 27, 2016): 336ed5, doi: 10.1126/scitranslmed.aaf5003.

229 *Daniele Fanelli at METRICS has also suggested:* Daniele Fanelli, "Set Up a 'Self-Retraction' System for Honest Errors," *Nature* 531 (March 22, 2016): 415, doi:10.1038/531415a.

230 *Paul Knoepfler at UC-Davis writes:* The Niche (https://www.ipscell.com).

230 *A British organization, the Faculty of 1000:* See http://f1000research.com/channels/PRR.

231 *Ahmed Alkhateeb, a postdoc at Harvard Medical School:* Ahmed Alkhateeb, "Opinion: Reimagining the Paper," *Scientist*, May 2, 2016, http://www.the-scientist.com/?articles.view/articleNo/46007/title/Opinion--Reimagining-the-Paper.

232 *A year after the journal started posting:* Mallory C. Kidwell et al., "Badges to Acknowledge Open Practices: A Simple, Low-Cost, Effective Method for Increasing Transparency," ed. Malcolm R. Macleod, *PLOS Biology* 14, no. 5 (May 12, 2016): e1002456, doi:10.1371/journal.pbio.1002456.

233 *The ARRIVE guidelines, for example, provide:* Carol Kilkenny et al., "Improving Bioscience Research Reporting: The ARRIVE Guidelines for Reporting Animal Research," *PLOS Biology* 8, no. 6 (January 29, 2010): e1000412, doi:10.1371/journal.pbio.1000412.

233 *A survey of animal-research guidelines:* Valerie C. Henderson et al., "Threats to Validity in the Design and Conduct of Preclinical Efficacy Studies: A Systematic Review of Guidelines for In Vivo Animal Experiments," *PLOS Medicine* 10, no. 7 (2013), doi:10.1371/journal.pmed.1001489.

233 Nature, *for example, requires scientists:* "Enhancing Reproducibility," *Nature Methods* 10, no. 5 (May 2013): 367, http://dx.doi.org/10.1038/nmeth.2471.

235 *"As with every economy, you may need":* Alex John London and Jonathan Kimmelman, "Why Clinical Translation Cannot Succeed Without Failure," *eLife* 4 (2015): 1–5, doi:10.7554/eLife.12844.

235 *It seems extraordinarily unlikely:* Daniel Sarewitz, "The Pressure to Publish Pushes Down Quality," *Nature* 533, no. 7602 (2016): 147–147, doi:10.1038/533147a.

INDEX